Hear the Ocean Sing

Published by The Lakehouse Publications
Lake Macquarie City, Australia
October 2019.

Copyright © 2019 Jan Mitchell

The rights of Jan Mitchell as author of this work have been asserted. Apart from use as permitted under the Australian Copyright Act 1968, no part of this book may be reproduced by any means without prior written permission from the author.

ISBN-13: 9780648497608

 A catalogue record for this book is available from the National Library of Australia

Printed in Australia in Book Antiqua font pt 10.

Author's contact: jan.mitchell2021@gmail.com
Web: writingsfromjanmitchell.com
Titles by this author available from the publisher:
10 Rosemary Row, Rathmines, NSW 2283

Also available from ingramspark.com, amazon.com.au and other online sources.

Hear the Ocean Sing

Part three of a cruising memoir

Jan Mitchell

The Lakehouse Publications

By the same author

tinker, tailor, soldier, sailor…the life of Colin Kerby OAM

Parts 1 and 2 of a cruising memoir:
Two in a Top Hat - A circumnavigation in *Caprice*
Crossings in Realitas

Frontispiece

Osprey A *with her new double-spreader mast, 2007*

Underwater Shape of the Brolga 33

Osprey A *on the hardstand at Marmong Point Marina, 2008*
Rudder and propeller removed for maintenance

Acknowledgements

The following people have been enormously helpful to me over the past several years during which I have written this book: my husband Ian, in correcting my mistakes about nautical technicalities and generally being encouraging to me in my writing activities; the Lake Macquarie Branch of the Fellowship of Australian Writers, who provide friendship, encouragement, learning experiences and critique; Alison Ferguson, Al Herring and Dirk Visman for reading the manuscript and pointing out errors; Frances Robertson for the cover design; technical support from Frances Robertson, Judy Handlinger and Nick Handlinger. Thanks also to Bruce Walker, friend, fellow sailor and reviewer of sailing books for his positive reviews.

Without my support team, this writer would certainly have been left floundering.

Unless otherwise acknowledged, all photographs were taken by me.

Table of Contents

Illustrations and Maps		i
Specifications of the Brolga 33		iv
Osprey A layout		v
About the author		vi
Boats we have owned		viii
Glossary		ix
Chapter 1	The Impossible Dream	1
Chapter 2	To Darwin with *Libelle*	19
Chapter 3	Finding *Osprey A*	43
Chapter 4	South to Tasmania	65
Chapter 5	The Tasmanian Wilderness	79
Chapter 6	Disaster in Bass Strait	105
Chapter 7	The Aftermath	123
Chapter 8	Voyage to New Caledonia	133
Chapter 9	Circumnavigation of Grande Terre	151
Chapter 10	Up the Clarence River	177
Chapter 11	New Zealand's East Coast to Stewart Island	185
Chapter 12	Returning North	207
Chapter 13	Lord Howe Island – again	227

Illustrations and Maps

1 *Osprey A* (2007) Frontispiece
2 *Osprey* at Marmong Point Marina

3	Fig 1. Interior layout of *Osprey A*	v
4	Jan and Ian Mitchell 2018	iv
5	*Realitas* under sail in Sydney Harbour	2
6	*Libelle* anchored in the River Derwent, Hobart	5
7	*Libelle's* saloon	5
8	Ian at the helm as we leave Hobart	9
9	Ian in *Libelle's* cockpit	12
10	Presenting Col with his biography on his 90th	16
11	*Libelle's* original galley	19
12	*Libelle's* solar array	20
13	Tim enjoying sunset, Graham's Creek, Curtis Island	25
14	*Libelle* anchored at Scawfell, Whitsundays	25
15	Jan stands at Cape York	27
16	Gove Aluminium Refinery	31
17	Deserted settlement, Port Essington, Victoria NT	33
18	The marina in Darwin	34
19	Judy and Jan cooling off in Roper River, Mataranka	36
20	Brolga in Katherine Caravan Park	36
21	Ian beside Landcruiser, westernmost point WA	38
22	Wildflowers in WA	39
23	Shifting sand dune, Eyre Bird Sanctuary, Nullabor	40
24	Our first view of *Osprey* at Scarborough Marina	43
25	*Osprey's* galley	45
26	*Osprey* on the hardstand at Marmong Point Marina	47
27	*Osprey's* deck is repaired and repainted	55
28	Jan repainting *Osprey's* cockpit	55
29	One of the ships scuttled for a reef at Tangalooma	58
30	A work in progress: David's houseboat interior	61
31	David on board his houseboat	61
32	Wineglass Bay from the shore	69
33	Looking back at the Denison Bridge	70
34	The Iron Pot, Storm Bay	70
35	*Osprey* moored at Constitution Dock, Hobart	71
36	*Swanhaven III* at anchor while we used their jetty	73
37	Whale sculpture at Recherche Bay	75

38	The rocky entrance to Port Davey and Payne Bay	78
39	Reflections in the Davey River	80
40	*Osprey, Dovetail* & *Taipan* moored at Kings Jetty	83
41	View from top of Mt. Beattie	84
42	Spain Bay looking inland	87
44	*Fig 2.* Chart of Macquarie Harbour	90
43	Hell's Gates: Entrance to Macquarie Harbour	91
45	Ian checks out a cell on Sarah Island	92
46	*Osprey* at Warners Landing on the Gordon River	93
47	*Storm Breaker* returns with rafters from St John Falls	96
48	Lake Fidler: a fragile meromictic lake	97
49	Early morning mist on the Gordon River near Eagle Creek	98
50	*Osprey* moored under the 'Nut' at Stanley	106
51	Jamie and Laura Anthony 2005	110
52	The Mitchells at the Maronite Church	111
53	View west from Mt Strzylecki (Flinders Island)	113
54	The foot of the mast is out of the mast step	115
55	A flying fire extinguisher hit Ian, blacking his eye	116
56	The targa arch is mangled	117
57	Dismasted 11.05.2005	117
58	The riggers raise the mast	128
61	The new targa arch on *Osprey*'s stern	130
59	*Osprey* is finally repaired and has her new rig	131
60	We set the worn MPS David had given us	138
62	Approaching the harbour entrance of Nouméa	140
63	View of harbour from above the cathedral, Nouméa	141
64	View from top of Amédée lighthouse	145
65	The old French Prison, Ile des Pin	147
66	Jean Marie Thibault Cultural Centre	148
67	One style of traditional native house	148
68	Lake Yaté water supply for southern town	149
69	Map of New Caledonia	151
70	Reef corals	152
71	The sea snake Ian caught in the galley	153
72	A two-palm tree islet in the lagoon near Koumac	155
73	*Fig 3.* Monument to Kermadec	158
74	Ballade Church font made from a clamshell	159
75	La Poule – the Hen, outside Hienghène	162
76	Jan with the coral trout she caught	164
77	Ian with four young Kanak men from Lavaissière	165
78	Marie Jo and Kelsey who led us to the waterfall	169
79	View from top of waterfall – *Osprey* at anchor	169
80	The weather turned dull at Woody Bay	170

81	Cagou – endangered national bird of New Caledonia	172
82	*Fig 4.* Map of Forster & Wallis Lake, NSW	178
83	*Fig 5.* Map of Clarence River, Northern NSW	181
84	*Osprey* departing the Clarence River	183
85	*Fig 6.* Map of Marlborough Sounds	189
86	My youngest brother, Greg Hormann	189
87	*Fig 7.* Map of NZ including Stewart Island (inset)	193
88	Seaward view of entrance Paterson Inlet	199
89	Weka on the beach, Ulver Island	199
90	*Osprey* anchored off Ulver Island, Paterson Inlet	201
91	The diving cage for viewing white pointer sharks	201
92	Jan outside the Parks & Wildlife hut	206
93	Otago Yacht Club	207
94	Our grandson, Tasman, at 16 months	208
95	The river port entrance at Whakatane	213
96	Auckland Harbour from Waiheke Island	214
97	A shallow anchorage on Waiheke Island	215
99	Jamie and Lisa are married, San Blas Islands	228
99	The Mitchell family at the summit of Mt Gower	231
100	A Catalina crashed on this historic site in 1948	232
101	Tropicbird on its nest, Mt. Lidgbird	236
102	View of Lord Howe Lagoon from Mt Lidgbird	238
103	Jan snorkelling	239

Specifications of the Brolga 33

LOA	33' or 10.16m
Beam	10'17" or 3.10m
Draft	6'82" or 2.08m
Displacement	15.284lb or 6.993kg.
Construction	Fibreglass
Designer	Joubert (Melbourne)
Hull Design	Long fin keel with skeg-hung rudder
Builder	Geoff Baker, Fibreglass Yachts, Mona Vale
First built	1965
Last built	by Formit Fibreglass in 1978
Osprey A	Built 1969 by Geoff Baker
Rig	Masthead sloop or cutter

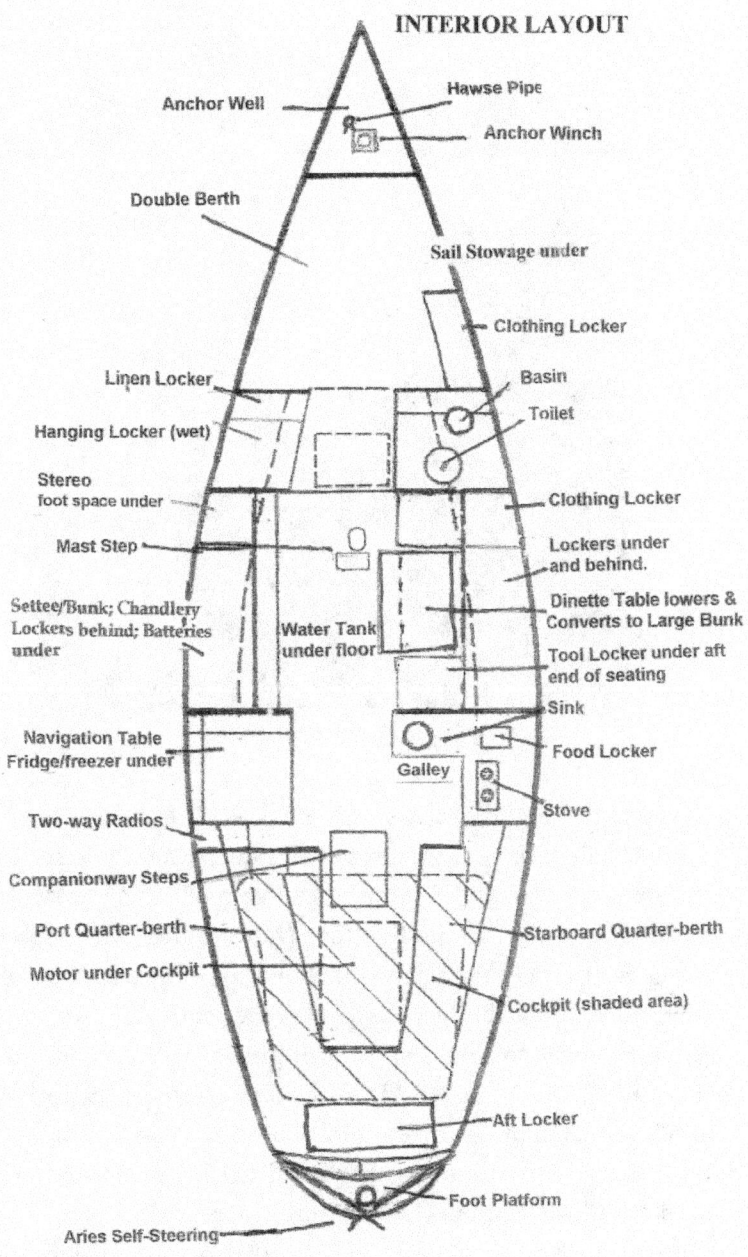

Fig 1. Interior layout of Osprey A

About the author

Ian and Jan Mitchell, 2018

Ian and I bought our first yacht in 1972, a steel Temptress design called *Jenny II*. She provided a steep learning experience, but was never going to fulfil Ian's dream of sailing around the world. *Caprice*, a twenty-five foot (7.6m) fibreglass sloop (Top Hat design) was a more realistic choice for inexperienced sailors with a limited budget. I wrote the story of our world circumnavigation in *Caprice,* originally publishing it in instalments in the bi-weekly magazine, *Australian Seaspray.* In 2012, I collated these articles into a book, *Two in a Top Hat.*

We returned from around the world with two-year-old Jamie, and David was born in Hornsby three months later.

To educate our boys and afford another ocean cruising yacht, we lived and worked for many years in Sydney.

In 1984, *Realitas*, a thirty-two foot Phantom yacht, became our holiday reality. Our family sailed up and down the NSW coast in *Realitas*, out to Lord Howe Island, which became a favourite destination, to Tasmania and to New Zealand. These holiday voyages developed in our two sons a love of sailing and the sea.

As soon as they had earned enough money, Jamie and David each bought their own yachts – Jamie a sister ship to *Caprice*, while David has concentrated on multihulls. This left Ian and me free in the new millennium to further our own cruising plans and we made a voyage to Far North Queensland in *Realitas*, before setting out to buy the yacht of our dreams.

We chose to buy *Libelle*, a forty-foot, thirteen-tonne steel vessel which proved too big and heavy for us to manage safely. A long search Australia-wide finally resulted in our purchase of *Osprey A*, a thirty-three foot fibreglass Brolga designed by Peter Joubert. We had the right boat at last.

Boats we have owned

Ian and I have owned a total of six yachts, all mono-hulls. Ian had dreamed of boats and sailing since late childhood. He inspired me and sailing became a shared goal. We married towards the end of 1971 and early the next year, bought our first yacht. Each of our boats has provided a learning experience, but perhaps none more than *Caprice*.

Jenny II	Steel Temptress (33')	1972 – 1973
Caprice	Fibreglass Top Hat (25')	1973 – 1978
Taria	Fibreglass Hood (23')	1981 – 1983
Realitas	Fibreglass Phantom (32')	1984.-.2000
Libelle	Steel Alan Payne P400 (40')	2000 – 2001
Osprey A	Fibreglass Brolga (33')	2004 -

Jenny II taught us that we couldn't hold down two full time jobs and make a steel yacht ready for a long voyage.

Caprice taught us to sail, and to trust her seaworthy design.

Taria became a stop-gap vessel after our world circumnavigation until we could afford another sea-going yacht.

Realitas was the ocean-going yacht that became our family floating holiday home for seventeen years. On *Realitas*, our boys learned to love ocean sailing.

Libelle, a beautifully designed steel yacht, proved too heavy for our age and health.

Osprey A turned out to be the closest to our ideal yacht – fibreglass for easy maintenance, well-designed by Australian engineer, Peter Joubert, and an excellent sea boat. She is just the right size and we still have her in 2019.

Glossary

Aft	Towards the back of the boat.
Antifoul	Paint to reduce marine growth on the hull.
Astern	Moving backwards.
Athwartships	Across the width of the vessel.
Beam	The width of the boat at its widest.
Berth	1. Bunk or bed. 2. Space alongside a wharf or in a marina.
Bilge	The lowest place inside the bottom of the boat.
Block	A single or multiple pulley with one or more sheaves enclosed between cheeks.
Boatswain's chair	Small temporary seat which is used to haul a person up the mast, e.g. to make repairs. Pronounced 'Bosun'.
Boom	A horizontal spar that holds the bottom of the mainsail.
Boom brake	A mechanism that controls the boom from swinging.
Bow	Front end of the boat.
Centre of gravity	Location where the total weight of the boat is said to act through, and where all motion pivots, so is the area with the least motion.
Close-hauled	With the sails pulled in tightly so the yacht sails to windward.
Coaming	The raised edge along the outer side of the deck, which restricts water entering the cockpit and that prevents items from rolling over-board.
Cockpit	The recessed section at the back of the boat from which it is steered.
Companionway	The entrance from the cockpit to the interior.
Danbuoy	Floating marker pole carried on-board and used to throw into the water to mark location of man or object fallen over-board.
Forepeak	Inside the boat at the bow, or front portion of the boat.
Furler	Device used to vertically roll-up the headsail to reduce (i.e. 'reef') sail area, or roll the sail up.

Galley	Area for cooking and food preparation.
Genoa	Large headsail used for sailing in light to moderate winds. Larger than a 'Jib' headsail.
GPS	The Global Positioning System by which the navigator uses satellites to determine the position of the boat.
Crosstree	Horizontal spar on the mast separating the rigging wires.
Gybe	To steer a boat so that a wind from astern swings the sails from one side of the boat to the other.
Gunwale	The outer edge of the deck, which is pronounced 'gunnel'.
Halyard	A rope used for hauling a sail up the mast.
Hard on the wind	The bow is pointed as closely into the wind as possible so that the boat sails forward. (NB: a sailboat cannot sail directly into the wind.)
Harness	Webbing worn by a crew member which secures him/her by a lanyard to a strong part of the boat, preventing a person falling overboard.
Heave-to	To slow or stop the boat by backing the headsail. Past tense – hove to.
Heel (of a vessel)	To lean to one side.
hPs	Hectopascals (units of air pressure).
Jib	Triangular-shaped headsail. A working sail for average conditions.
Keel	Heavy section of the boat under the hull that works to keep the boat upright.
Knots	A knot is one nautical mile per hour. Boat speed is measured in knots.
Lanyard	Rope used to secure (e.g. a person) to a strong point.
Leads	Markers to assist navigation into an enclosed area.
Lee cloth	A piece of canvas that is firmly attached to the "lee side" of an open-sided bunk. It keeps the sleeper from falling out when the boat heels.
Lee shore	The shore (land) downwind of the boat.
Limber holes	Holes used to drain water from one area to another.

Lines	Ropes or lines on a boat have different names according to their purpose.
Log	Mechanical or electronic device for measuring boat speed through the water.
Nautical mile (nm)	Distance as measured over the surface of water. One nautical mile equals 1.852 km or 1.151 statute (land) miles.
Nav. Station	Area where the navigator works out calculations to determine the boat's position, marks the charts and keeps official records of the voyage.
Navigator	The crew person who calculates position of boat and records its progress. He/she also uses the radios to communicate with other boats and shore stations.
Outboard	Small self-contained motor used to propel a dinghy or other vessel.
Painter	A rope attached to the bow of a dinghy which is used to tie it up.
Port	The left hand side of the boat, when facing forward.
POB	Abbreviation for 'Persons on board'.
Practique	Process of going through customs and immigration when a boat enters a foreign country.
Quarantine	All crew on a boat must remain aboard until cleared by a health officer. A 'Q' flag is flown from the mast to indicate a health clearance is required.
Quarter berth	Bunk (bed) located to one side and partly under the cockpit.
Radio channels	Dedicated radio frequencies for specific communication purposes, such as universal calling and emergency channels.
Reefing	Reducing the sail area by tying up or furling part of the sail
Rudder	Underwater mechanism to steer the boat.
Self-steering	An electronic or mechanical system capable of automatically steering the boat in a set direction.
Skeg	A projection behind the keel to support the rudder.

Sheave	A grooved wheel which directs or changes the direction of a rope or line.
Sheets	Ropes connected to the lower back corner of a sail (the clew). They are used to tighten or loosen the sail and control the amount of power generated by the sail.
Shrouds	Rigging wires attached from the cross-trees to the deck.
Spinnaker	Very large, often colourful, down-wind sail flown from the front of the boat.
Spinnaker Pole	Long, light spar used to hold the spinnaker or genoa in position.
Starboard	The right hand side of the boat, when facing forward.
Stays & Shrouds	Strong wire or rope rigging used to support a mast (e.g. the 'forestay' runs from the top of the mast to the bow).
Stern	Back of the boat.
Stringer	Longitudinal timber which reinforces the hull.
Tack	To turn the boat across the oncoming wind so that the wind now comes into the sails from the opposite side of the boat. (A sailboat must sail a zig-zag course to make way into the wind.)
Tinny	Slang for aluminium dinghy.
Trade winds	Winds that blow regularly in the tropics – NE in the northern hemisphere and SE in the southern hemisphere.
Turning block	A device containing a grooved sheave or pulley which is used to change the direction of force on a line (rope).
VHF & HF radio	Two-way radios for communicating ship to ship or ship to shore. They have multiple communication channels as well as dedicated calling channels for different purposes. VHF signals travel by line of sight and HF has a much greater range of signal.
Yankee	A high-cut foresail.

Windvane	Literally the vane attached to the self-steering system, which is moved by the wind, and controls the mechanism. The term is commonly used to mean the entire mechanical self-steering system.
Wing out	To position one sail either side of the boat, like wings, for running downwind. Past tense – wung out.

1 The Impossible Dream

The ocean is a mysterious mistress. She bewitches, she beguiles, she lulls, and then she envelops, threatens and displays all her might – often in the middle of the night. The ocean sings in all the moods expressed in an opera. And sailing across oceans is a strange and extraordinary experience – one I have made in company with my husband, Ian, for many wonderful years and in which I can no longer participate. Instead, I explore our past adventures in my writing.

I had learned to snorkel during our voyages in *Realitas* in the 1990s and I wanted to continue to observe the fabulous underwater coral gardens I now loved. I was disappointed we had sailed right through the West Indies during the mid-1970s and I had no idea of the wonders lying beneath the sea surface. I had discovered these delights on a visit to Lord Howe Island in *Realitas*. Now I wanted more voyaging to exotic places and coral islands.

'I think we should sell the house and live on board the boat,' said Ian.

'We need to keep some equity in the Sydney market. How about we buy a home unit and rent it out? That would give us income, too.'

'That's what I was thinking. Something where there's no garden to upkeep and little maintenance.' I said. 'And it's time to start searching for our next boat. If we're going to live aboard, we need something more comfortable than *Realitas*.

'I suppose so.'

'We need the new boat first, so that we can move on board as soon as the house is sold.'

Over the previous seventeen years our thirty-two foot Phantom yacht had proved a wonderful choice for a holiday boat while our two boys were growing up. Now it was just Ian and me again. The previous year, we'd sailed *Realitas* up to Far North Queensland and returned in late November (see *Crossings in Realitas*).

Successfully sailing so far north had shown us that we could manage long distance cruising again and that being at sea did wonders for Ian's health. He was looking and feeling the best he had in the five years since he'd been diagnosed with Chronic Fatigue Syndrome.

We scoured the newspaper columns of boat advertisements for a suitable yacht. Often, we went out to look at boats that initially seemed hopeful, but on seeing them, rejected them all.

Realitas *under sail in Sydney Harbour*

One of the most interesting boats we looked at was an Alan Payne Tasman Seabird. Alan Payne's designs had long fascinated Ian. I remembered from years before sailing into Refuge Bay in the Broken Bay area north of Sydney, Ian had excitedly pointed to a multi-chine steel boat and almost breathlessly told me, 'That's an Alan Payne!'

The vessel was painted yellow and I thought it looked like a steel banana on the water. For some time after, I referred to the Seabirds as 'banana boats'.

I gave close scrutiny to the potential stowage on the Seabird.

'This boat is 37 feet long, yet it's not as roomy as *Realitas*. I'm not interested in a boat with less storage than we already have,' I told my husband.

I was looking for more physical comfort, hence a slightly bigger and heavier yacht than the Phantom 32, which the Seabird was. Although it sailed more smoothly than *Realitas*, I also wanted more stowage for long passages and a fridge/freezer for Ian's changing dietary needs. Besides, to me, that Seabird didn't feel welcoming and homely.

One day, we saw an advertisement for a yacht in Hobart. It was a forty-foot Alan Payne design. It sounded like the kind of boat that would leave the Chilean channels in its wake. We were both entranced with photographs we'd seen of the Chilean channels and I still dreamed of sailing there. If I was keen, Ian was too. Maybe we could meet both our dreams with this boat in Hobart.

When I had decided to throw my life in with Ian as his partner and wife, I had seen part of my role to promote into reality things he dreamed about. Sometimes Ian lacked the will to make his dreams happen. I have a greater degree of self-assertion than he has and I was willing to use it to create a happier life for my beloved husband, friend, and partner in life's adventures.

'Let's put an offer on it,' I suggested.

Ian was enthused by the advertisement and without even having seen the boat, he phoned the broker to make an offer $10,000 lower than the asking price.

A couple of days later, Ian's phone rang while we were attending an appointment.

In the car on the way home, Ian told me the yacht owners were interested in our offer and I was elated.

'We'd better book ourselves plane tickets to Hobart,' I said.

'I need to think about it all a bit more,' my husband demurred.

'Why? You've admired Alan Payne's designs for so many years. This is the sleek, long-keeled, steel yacht you've always wanted.'

'It's a big boat. Bigger than we had agreed upon.'

'I know. But we need more on-board stowage than *Realitas*. The Seabird didn't have it and nothing else we've looked at seems suitable.'

Ian had seemed keen when he saw the advertisement. Now I wondered why the possible ownership of this solid steel vessel didn't seem so appealing. I thought I knew the reason.

'Don't you think you deserve such an expensive boat?' I said. He looked thoughtful. 'We can live permanently on a forty-footer. You are always happier and healthier living on a boat.'

'Yes, but is this the right boat?'

'We won't know until we go down and have a look at her. They say she is ready to cruise the world. We need to check her out.'

Ian still looked doubtful.

'Well, while you're thinking about it, I'll see what flights are available to Hobart.'

Here was a boat that could take us anywhere, in just about any conditions. I wasn't going to pass up an opportunity to look at her.

In Hobart, we settled into the small holiday apartment I'd booked for a couple of days, and then headed down to the waterfront to inspect the yacht.

A very dark green boat was tied alongside the wharf. The name printed on her bow was *Libelle* - not an altogether unlikeable name.

She looked every bit the strong steel cruising boat we were looking for, though I was surprised at her colour. The very dark green paint made her hull look almost black. In my opinion, an orange colour on a steel hull both hides rust stains and makes for a more visible vessel at sea.

The agent took us on board. I gasped at the light lettuce-green of the interior. Expensive plain woollen upholstery covered large surfaces, including the double bunk/settee on the port-hand side of the saloon. I might as well have been worming my way inside an iceberg lettuce. It made me feel seasick just to look at it.

Oh well, the colour could be modified, I thought. Cushions and throw rugs would all help soften the eye–smack.

'How about you check the rigging and deck hardware while I look over the galley and sleeping quarters?' I said.

Ian went on deck and checked the vessel over with his usual meticulousness, testing lights, winches and rigging, then came inside, looking under floorboards, inspecting the engine and more. I looked inside lockers, tried the bunks, looked for and found lee cloths, and inspected the galley. We both looked for signs of rust. The yacht appeared to have been very well maintained.

'At least she has heaps of stowage, especially under the floor,' I said.

Libelle *anchored in the River Derwent, Hobart*

Libelle's *saloon*

'We need some time alone to think things over. We'll come back in the morning,' Ian told the broker after we'd been on board for over an hour. We set a time to return the next morning and walked from the water-front back to West Hobart, to our holiday flat. On our walk, Ian seemed enthusiastic. He talked excitedly about the yacht and I assumed he'd made up his mind that this was the boat for us.

I was more concerned about the asking price. It was really stretching our financial limits to spend that much, especially on a vessel which would demand greater maintenance costs than we were accustomed to. And we needed to sell *Realitas*. However, I was on a high. The contemplation of purchasing such a large-ticket item was very stimulating to my bi-polar self.

'You know, they advertised *Libelle* as being ready to ocean cruise,' I said to Ian. 'She's not really ready. The water tanks are full of aluminium flakes, the life raft is out of service and probably needs replacing and the HF radio needs up-grading. She also needs a certificate for the gas fittings, and more.'

'You're right. There are quite a few things that would need to be done before we could go any real distance in her.'

'We could use those to bargain for a price reduction.'

'I don't know if I would be able to do that,' said Ian.

'I could. How about you let me do the negotiating? To have a reversal of roles would really throw them.'

'Hmm. That might work to our advantage. Are you sure you can manage to carry it off?'

'I think we should go for it,' I said.

'I don't know,' said Ian.

'It's the sort of boat you've always wanted. We've got the time now to look after steel.'

'She's big. I could go running around the deck.'

'You could.'

'If we could get her for an affordable price, would you go for it?'

'I might.'

'I hardly slept last night, Jan,' Ian said in the morning. 'I'm not sure we're doing the right thing here.'

'I think you believe you don't deserve this boat,' I replied. 'You are worth it. You should follow your dream now we can afford to do it.'

I wasn't going to be talked out of negotiating for *Libelle*. We walked down to the broker's office on the waterfront. As I predicted, they were taken aback by the change in tactics. I hammered home what it would cost us to bring *Libelle* up to scratch for a long voyage. I was hoping we could go to Chile in her, via the Southern Ocean. That was the dream anyway.

In the end, I beat them down to the price we could afford to pay. I was proud of myself and somewhat elated. This is a bad sign in a person with bi-polar disorder, but at the time I was very good at hiding recognition of my mood. I should have realised I was being irrational, because that night Ian again found it difficult to sleep. As is normal in boat purchase agreements, the settlement is conditional on a satisfactory test sail and an out-of-the-water inspection.

Typical of Hobart, while we were on the water the sky clouded up and the wind increased. The weather can change completely in less than an hour. Even when the wind gusted close to thirty knots (about 50kmh) this yacht was so well balanced that she was easy to steer.

'Jan, come and take the tiller from me,' Ian called.

I did and was amazed. It required no effort at all to steer such a big, heavy boat.

'Wow, compared to *Realitas*, this is a breeze,' I said.

Having seen and felt enough, we agreed to proceed with the out-of-water inspection. The broker had made slipping arrangements across the River Derwent at Bellerive and found an inspector for us.

After lunch, Ian and I went by bus across to the boatyard where *Libelle* was being winched up the rails. She was as good out of the water as she had looked in her element. The inspector could find no sign of rust – the steel hull had been properly maintained. There were signs of early corrosion that we'd already noted at one spot on deck where some extra welding had been done to encourage any water to run off towards the gunwales. This didn't look too difficult to keep under control. I convinced myself that in retirement we could cope with looking after steel. After all, we were planning to live on board this big forty-footer. There would be no travelling back and forth to the boat. We and the tools would be right there.

I was elated. How typical of me when I was feeling manically optimistic. Leaving *Libelle* on the slipway, we found the bus from

Bellerive back to central Hobart. We sat on the waterfront and ate fresh oysters out of their shells by way of celebration.

Having signed all the paperwork and paid our deposit, we flew home. Ian was still full of misgivings but I was satisfied that I had managed to buy my husband his dream yacht. I ignored the fact that he'd had sleepless nights before we committed ourselves. I was hyper on the belief that, once again, I had helped achieve Ian's dream for him.

It took several months to find a buyer for *Realitas* and, for a short time, we were the owners of two yachts. The house was a different matter, though. The real estate agent found us a buyer almost immediately he'd hammered the 'For Sale' sign into the grass outside our front fence. Friends of some Chinese neighbours rang the agent that very evening with their first offer. Being new immigrants, they were determined to buy in our street close to their friends. They made an attractive offer and we accepted.

We needed to bring *Libelle* home to Sydney harbour so that we could move on board properly and place into storage the furniture and other belongings we wanted to keep. Life moved at an accelerated pace. There was lots of cleaning and packing to be done, selling of unwanted items and innumerable other jobs to be attended to. Eventually we were ready to fly back to Hobart and bring *Libelle* home.

When she was fully ours, I filled up the lockers with food, while Ian checked out the engine, topped up the fuel tanks, renewed the water – and finally, we set sail for Sydney.

The weather was favourable to start with. By the time we were in Bass Strait, there was a minor front predicted to pass through. We hove to for a few hours until it had passed, and then made sail again.

I was delighted to find that the steady motion of this boat made it pleasurably comfortable to walk around or lie down and sleep. Unlike our earlier boats, we were not bounced about. She was heavy and stable. The image of a vessel sailing through whipped cream came to my mind and has stuck there.

'Look Ian, no bruises,' I said, showing him my legs. Normally after a cruise, my legs and hips were covered in bruises.

'Hey, that's good. I'm not bruised either.'

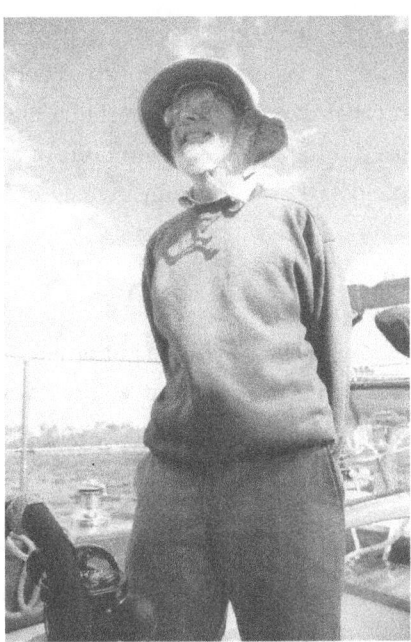
Ian at the helm as we leave Hobart

We brought *Libelle* into Five Dock Bay, but we couldn't use our mooring because *Realitas* was still on it. We decided to anchor as close as we could to the moorings, without endangering any boats when the wind changed. This meant we were virtually in the channel leading to the end of the bay.

'There are only moorings up there,' I said. 'It's not as if there's a wharf. It shouldn't bother anyone to leave her here.'

'I hope you're right.'

I wasn't right. Someone complained and two weeks later, we found the Waterways officers had tied a 'Please Remove this Vessel' notice to our boat. Before they could tow *Libelle* away, Ian motored her to Balls Head Bay, while I drove the car there. We anchored her in the bay that had been Jamie's home on *Possibilities* for his three undergraduate years at Macquarie University, and then returned home to complete the move out of our house. Balls Head Bay has park surrounding it and no overlooking houses with people who might complain about live-aboard yachties. In fact, this was the bay where Waterways advised visitors to Sydney to anchor their boats. We thought *Libelle* would be safe from harassment there.

The house sale went through two days after being advertised. I organised a garage sale for items we no longer needed, advertising the sale to begin at 8.00 on Saturday morning. By 11.00 the night before, I was exhausted and collapsed into bed, leaving an hour or so of work still to be completed. At 6.00 o'clock, I was awoken by a loud hammering on the front door.

'You advertised a garage sale?'

'Yes, but it doesn't open for two hours yet.'

'I won't return. Show me what you've got now.'

Reluctantly, I opened up the back flat where I was holding the sale. Fully awake by now, I realised these people were secondhand dealers. More arrived. I barely had a chance to get dressed. Ian didn't help. The sale was my thing.

By 11.00am, my head was spinning. People were trying to take garden tools for fifty cents each that new had cost tens of dollars! Eventually, no more people came and I was left with the rubbish. I felt like rubbish too. My nerves were shattered and I was helpless. I needed sustenance; I needed peace, quiet and rest; but I was beyond rest and I couldn't sleep. I was falling over the edge into the pit of depression. This time, the pit didn't approach over the next several days. I merely sank straight into deep darkness and curled up.

Moving day was fast approaching. As sick as I was, I pushed on, both of us packing as best we could. We were storing our belongings in the double garage which belonged with the unit we'd purchased in Hornsby. The house sale and the unit purchase were completed simultaneously and we rented out the unit without any car space.

When our younger son, David, had finished his first semester exams at the University of Newcastle, he came home to help us with the move. The truck arrived and Ian and David helped the two removalists trolley the boxes down the front steps while I huddled on a concrete step at the back of the garden. There my neighbour found me that chilly grey day in May. She coaxed me into her kitchen to sit by her heater and gave me a hot drink. Good neighbours are a godsend.

We drove back down to Wollstonecraft, where we'd anchored *Libelle*. On board, we managed to get some food and go to bed. I awoke early the next morning and looked around the boat.

With a sinking heart I said to Ian, 'I've made the most enormous mistake. We shouldn't have bought this boat. Why didn't you stop me?'

'How could I stop you? You were like a steamroller.'

I turned over in bed and put my head under the covers. Shortly after, I could hear the motor of a vessel alongside. A voice hailed us. I staggered out of bed and dragging on a jumper, climbed into the cockpit. It was the Waterways Officer.

'How long are you intending to stay here?' he said. 'You can't stay here. You'll have to move.'

'I'm sick,' I said and sank down onto the cockpit floor.

He couldn't cope with that and went away. He returned the next week at the same time. This time, Ian went out.

'How's your wife? Is she better?' the Waterways Officer asked.

'No, she's still sick,' said Ian.

'I'm sorry, but I'm afraid you can't stay on your boat here. You'll have to find a marina berth.'

'Okay.' Ian never argued with authority figures.

When he'd gone, we talked about what we could do. I was so depressed, I really didn't care. We couldn't stay on *Realitas* on her mooring, because we needed to keep her looking pristine for a potential buyer. And there was one coming up from Victoria in a few days.

'How about we go up to Lake Macquarie?' said Ian.

'I suppose we could. I'm not going to be much help though.'

'We'll manage somehow. We'll just take it easy. 'We can leave the car parked here and come back for it by train.'

Ian was being the planner – usually my role. I was incapable of making decisions in this mental state.

Ian moved Libelle further down the harbour to our CCCA (Coastal Cruising Club of Australia) mooring at Bottle and Glass Bay, just near Vaucluse in the Eastern suburbs. We stayed there for a couple of nights, waiting for a favourable wind to sail north. The first day, we sailed as far as Pittwater, stopping at the Basin for a night on the club mooring there before continuing on. I slept while Ian sailed the boat.

Fortunately, the conditions were gentle. At Moon Island, he turned *Libelle* towards the land. After entering Swansea Channel and passing under the Pacific Highway Bridge, we waited on a mooring for high tide before motoring through the dog-leg passage into Lake Macquarie. We bumped a couple of times as we came across the drop-over, and then we were into deep water again, well away from that pesky Waterways Officer.

Ian had booked us into a berth at Marmong Marina, near the head of the lake. He signed the paperwork saying we were aware of the rule that no one should live aboard, and we settled into the marina along with the other live-aboard people. Now we had hot showers and toilets available just down the pier.

Ian in Libelle's *cockpit*

I didn't want to be left alone, so we travelled together by train back to Sydney. The car was just where we'd left it, and by the end of the day she was secure in the locked carpark used by the Marmong Marina berth holders. Having somewhere safe to stay was our first 'win' since moving out of the house.

A few days later, the broker in Sydney rang with an offer for *Realitas*. Maybe we quibbled and the buyer offered a little more, but I don't remember. We were just pleased the agent had found a buyer, although we were a little surprised to find that the new owner of the yacht, which had taken us on so many ocean voyages, intended to have her trucked to Victoria.

'Why is he buying an ocean-going vessel if he doesn't intend to sail her?' I asked. Ian shrugged.

'His business. Not your worry – nor mine.'

Ian was right and when the bank cleared the cheque for *Realitas*, we had funds again. We forgot about *Realitas* and concentrated on *Libelle*.

Gradually we began to know by face and then meet others who lived on their boats at Marmong. Paul and Helen had made a beautiful home afloat for themselves and their two children. Helen introduced us to Judy and Colin Kerby, who were to become very special friends. Colin was very amusing. If he met me on the jetty, he would do a little dance – very funny when he was still in his overalls and dirty from his day's work at a smelter. He loved to tell stories about his long and eventful life. We always enjoyed the moment when he said, 'Did I ever tell you about …'

Ian settled into planning what alterations we needed to make to *Libelle* before going cruising. Our very first priority was to do something about the water tanks. They were made of aluminium and corroding so badly that flakes of aluminium were coming through the galley tap. Even though we were filtering the water, it was far from a satisfactory arrangement. Our plan was to put in stainless steel tanks. A member of our yacht club could make them for us in his Sydney tank manufacturing business.

We puzzled about how to get the tanks out and new ones in. Ian thought he'd have to dismantle the companionway, yet he made no detailed measurements of the spaces. I found that strange, but was too depressed to query his assertion. Slowly the idea became fixed in his head that the companionway was too narrow for the tanks to pass through.

Ian was no longer clear-headed. He had experienced far too much stress over the past few months and his Chronic Fatigue Syndrome was becoming worse again. I thought that he, too, was sliding into depression, but it was chronic fatigue-induced brain fog. We were living on what I had thought would be his dream boat, but we were going nowhere and I certainly wasn't a cheerful living companion.

Neither of us could see a way out of our situation. We began to resent the boat. Ian probably resented me for talking him into it. He just didn't know where to start with maintenance tasks that should have been easy for an engineer with his skills. The brain fog was making decision-making difficult for him. As for me, I couldn't assist on *Libelle* as much as I had on either *Caprice* or *Realitas*. I tried to take the genoa from its stowage place up to the deck, but I couldn't even lift it. It was simply too heavy. I couldn't put the mainsail cover on because the top of the boom and the zipper on the cover were too high above my head. A ladder is not really appropriate on a moving yacht!

Inside my realm, the galley, I found it difficult to reach to the bottom of the dry goods locker I had nicknamed 'the cave'. When I spilled a container of rice, I couldn't reach to clean it up. Likewise, I couldn't reach the bottom of the fridge. The original owner, who'd fitted out *Libelle*, was a touch over six feet tall; I am five foot two and a half. (I always include the half because when you're short, every half-inch matters!) Much of the stowage was under the floor. It seemed I was on my knees far too often.

'If I'd wanted to get on my knees this often,' I said to myself, and later to Ian, 'I would have joined the Catholic Church and become a nun.'

We decided to put *Libelle* up for sale. She represented more work than we thought we could handle. A couple came to look at her. Like us, they wanted a retirement boat for cruising and decided they wanted to buy. We arranged for a deposit and also to take them out for a test sale.

The nominated day for the test sail was blustery. Despite our long years of cruising, we were very inexperienced at bringing any boat in and out of a marina berth. We should have postponed the test sail. We didn't. Trying to back *Libelle* out of the berth was a disaster. She weighs thirteen tons. When the wind caught her broadside, there was no way I could hold her. It was like trying to hold a rearing horse. Rather than be dragged into the water, I had to let the rope go. *Libelle* crashed into the neighbouring vessel. I nearly vomited with tension and Ian was near tears. During a lull in the wind, we managed to calm the horse back into her stable.

We inspected the damage. Incomprehensibly, there was barely a scratch on the boat we'd hit. Quite how that had been avoided, we didn't know, but we were grateful we didn't have to deal with an insurance claim. *Libelle* was not so lucky. Above her steel side the wide timber bulwark had cracked.

Our potential buyers went off to their motel, promising to return the next morning. Instead, they phoned from their car.

'We're already driving home. We've had a sleepless night. We've decided *Libelle* is not for us. Would you please return our deposit?'

Ian did. We were bitterly disappointed that the sale had fallen through and my depression deepened. To help us assuage our gloom, Judy and Col Kerby often invited us onto their boat, *Ooroo*, for afternoon tea. Occasionally, we reciprocated. Col had recently retired from work at the smelter and, nearing eighty, he was planning to prepare their boat for a voyage to Papua New Guinea.

Col is a wonderful raconteur and he told us many stories of his eventful life. His eyes would gleam as he began to speak and Judy would shake her head.

'No Col. Not that story.'

'Go on; tell us,' I would encourage him. And Col would tell the story anyway while Judy covered her eyes.

Sometimes, Judy would correct Col over a small detail. Up until that stage, we had been unsure whether these stories were true or not. However, they were highly amusing and we were happy to listen, because they were a distraction from our gloom.

One day, Judy went into their cabin and returned with an enormous ring binder. She flicked through the folder until she found the page she wanted and turned it around for us to see the newspaper cutting. We realised then that these stories of Colin's were all true and Judy had everything documented in several very large foolscap folders.

A few years later, I wrote Colin's biography and became very familiar with those detailed records of their lives. (See *tinker, tailor, soldier, sailor…the life of Colin Kerby OAM*)

In November 2011, on his ninetieth birthday, I attended Col's birthday dinner and there, I presented him with the first copy of the book.

My deep depression continued, as did Ian's chronic fatigue. We were still unable to make decisions. *Libelle* was still advertised for sale. David, who was living in nearby Newcastle, came to visit occasionally.

Presenting Col with his biography on his 90th Judy Kirby on R

In December, Tim Lamble entered our lives by answering our advertisement for *Libelle*. Tim had a very fixed idea about the forty-foot boat he should buy for his proposed circumnavigation. It would be designed by Alan Payne and would have a wheel for steering. Other than not having a wheel, *Libelle* fitted his dream.

We took Tim out sailing on Lake Macquarie, showing him how well-balanced the boat was and how easy to control with a tiller, how much easier and cheaper it is to set up an *Aries* wind steering system with a tiller. Ian explained the difficulties of maintaining a wheel steering system when lines run under the cockpit floor. It took Tim several visits to us before he reconciled himself that his dream of sailing around the world could include the yacht having a tiller.

During Tim's visits, we became good friends and our relationship with him lifted our mood.

As we moved into the New Year (2002), Tim made up his mind about purchasing *Libelle* and on the third of January, presented us with a cheque for the entire purchase price. *Libelle* became Tim's boat and we had but a week to find ourselves alternative accommodation. Once Tim had paid for *Libelle*, a thirteen tonne weight lifted from our shoulders. Ian became buoyant, and the worst of my depression disappeared overnight.

It was only five months since we'd let our unit in Hornsby. The fixed period of rental was for six months. We immediately gave the obligatory six weeks of notice to our tenants, but were left to find other accommodation until then for ourselves. Unfortunately, my sister was unable to accommodate us for a full six weeks. This was understandable, because her unit has only one bedroom and she needs lots of sleep. Also, her job requires her to be at work by 6.00am.

Members of the CCCA came to our rescue. Brian and Jill Robinson had retired to a village near Dora Creek. They kept their boat, *Narama*, on the Marmong marina and we saw Brian frequently when he came up to maintain his boat, a Joubert designed Brolga 33.

Brian and Jill offered to have us stay in their house for three weeks. This was a godsend, because January, being school holidays, is a difficult time to try to find short-term accommodation. Brian and Jill were spending two of those three weeks in Sydney. They were happy to leave us in charge of their house.

Morning walks around the village were very different. During those weeks, we sighted many white-haired people walking their little white dogs, but in the village we saw no children and no teenagers. In our late fifties, we were the youngest people around. It was an ideal short-term arrangement, but that taste of retirement village life convinced us we never wanted to live in a retirement village.

Besides, we had a lot more living to do yet. There was the matter of finding our next yacht, but that was to take us another couple of years.

2 To Darwin with Libelle

Tim and a friend of his moved *Libelle* from Lake Macquarie down to Pittwater and moored her at Mitchell's Marina at Church Point. Despite the name, this marina has nothing to do with our family, except that it was where we'd bought *Caprice*, way back in 1973. (See *Two in a Top Hat*.)

Tim completed a long list of tasks that modernized his yacht. First, he removed the mast and refurbished it and the rigging wires. Later, he and Ian replaced the aluminium water tanks with stainless steel, installed a wet locker, replaced the toilet, installed a holding tank for it, and also upgraded the electrical system with several solar panels, a wind generator and new absorbed glass matt batteries for storage. Ian happily caught himself up in *Libelle's* renovations. He was surprised to find that the companionway frame did not have to be removed to take out the water tanks. Later, Tim renovated the galley.

Libelle's *original galley*

After Tim sold his house in Turramurra, he began couchsurfing with friends until such time as *Libelle* was ready. Our friendship with Tim had developed to the point that our flat was one where he frequently stayed, and we were very comfortable with that.

Tim had worked as a draftsman for some big engineering corporations. He had also been a writer and editor for BHP's prospectuses and information leaflets. We had lots in common – engineering knowledge and writing experience and similar political attitudes, as well as an interest in ocean sailing. However, Tim's only ocean sailing experience had been to bring *Libelle* to Pittwater, approximately thirty nautical miles. Most of his sailing had been on Sydney Harbour on VJs and other small boats during his youth.

Libelle's *solar array*

Although Ian was busy with *Libelle*, we continued to look for another boat for ourselves. Ian reminded me how much he'd liked the Brolga design by Joubert, but we weren't happy with the interior plan of *Narama*, the Robinson's boat, which was for sale at Marmong. It had a central galley instead of a saloon. There was nowhere to sleep at the boat's centre of gravity. It was very important for me to be able to lie down when I was seasick.

I remembered a day long ago in 1978 when *Caprice* was on a mooring close to the Pittwater Sailing Club and Ian had pointed to a vessel on a cradle in the hard stand.

'That's a Brolga,' he'd said. 'It's designed by an engineer called Joubert.'

'His name sounds French.'

'No, he's an Australian engineering academic from Melbourne. I'm not sure if he works at Melbourne University or Monash. That boat would be ideal to sail around the world, but it's definitely out of our price range.'

Tim Lamble was friendly with Jim and Jenny Starling, who owned a Brolga 33 and the couple were members of the CCCA, our sailing club. Another connection was that our sons had been at high school with their son, Jeff. Maybe it was their yacht, *Virgo*, we'd seen that day as they kept their yacht at that same sailing club. Tim told them we had begun looking for a Brolga.

Jim and Jenny offered to take us for a sail on Pittwater on *Virgo*. Jim and Jeff frequently raced the boat but didn't cruise her very far, so *Virgo* was set up for very efficient sailing.

One lovely sunny autumn afternoon in 2003, Ian and I went out on Pittwater with Jenny and Jim. We were most impressed with *Virgo's* sailing qualities and her level of comfort when heeling to the wind. We could afford a Brolga now and we wanted one. Unfortunately, there were probably fewer than thirty ever built, so they didn't come onto the market very often. And the Starlings weren't offering to sell theirs.

I later found out that Joubert had designed three other moderately sized racing yachts also comfortable for cruising, plus some other vessels. His designs were created during the sixties and seventies when people who grew up after the Second World War were stretching the boundaries of conventional life and many set off in small yachts to sail around the world.

Apart from the 33' Brolga, there was a 24' Koala trailerable yacht, the 27' Pelican, the 31' Currawong, the 34' Magpie, the 37' Cape Barren Goose and the 43' Bounty. Some of Joubert's designs raced in the Sydney to Hobart yacht races. Peter Joubert died in Melbourne on 13 July, 2015, aged 91.

Ian and I were itching to go cruising again and felt we'd like to do it in a Joubert designed yacht. At that time, we'd been on a Magpie and a Cape Barren Goose and then the Starling's Brolga, as well as the one at Marmong that Brian Robinson owned - the Brolga with the galley amidships.

We were interested in the design with the galley near the cockpit and opposite the navigation station, but they were hard to find for sale.

In the meantime, we wanted to go sailing by whatever means. I saw our opportunity when the young man whom Tim had intended to take with him as crew pulled out.

'How would you feel about us volunteering to sail with Tim to Darwin?' I asked Ian.

'I think I would like that very much,' he said. 'I'm enjoying working on *Libelle* and I rather like the idea of sailing her when I don't have the ultimate responsibility. I think we could get along with Tim on a voyage.'

'It would be a very different experience being crew on a boat we've owned,' I said.

'Yes, but I think we could adapt. We both like Tim.'

At the next opportunity, I asked Tim, 'Would you be prepared to take Ian and me to Darwin as crew?' I always believe in being direct. I hope there isn't room for misinterpretation that way.

Tim wouldn't give an answer straight away. He needed to mull the idea over to see if it sat well with him. A couple of months later and after much badgering from me, he eventually agreed that we could sail with him, Ian as navigator and me as cook. We discussed terms. We would pay for all food during the trip north and Tim would pay for the fuel.

I began planning our side of the trip. Tim had decided to leave in April 2003 – just three and a half months away. Ian and he were very busy completing their list of tasks on *Libelle*. In particular, Tim was very concerned that all the electronics he wanted were properly installed – the latest in High Frequency (HF) two-way radio, an alarm system for the proximity of large ships and a computer system integrated with the GPS for navigation. He wanted these items to complement the radar that previous owner, Peter Hooks, had installed. Tim also bought a new *Canon* digital camera, which failed within a month of our departure.

Ian felt impatient with all this emphasis on electronics. We both knew how easily they could all fail in a damp marine environment or an electrical storm. But this was Tim's voyage on his yacht, and that was what he wanted. He added more solar panels to help run all this equipment.

Once we left Tim and *Libelle* in Darwin, Ian and I intended to buy a campervan and travel around the rest of the Australian coast. I was very excited as we made our plans.

'Judy and Mike want to drive up and join us. They want to see Darwin, Katherine and Kakadu.' Ian was thrilled to hear that our best friends wanted to come so far north from Tasmania.

'They probably won't join us in the Kimberley, though,' I added.

We planned to visit the Argyle Dam, drive through the Kimberley to see the Bungle Bungles, then return to the coast to Broome, Exeter and then south through Western Australia to see the famous wildflowers. We intended to be absent for at least six months.

'We'll need to buy a four-wheel drive. I suppose we'll be able to find one in Darwin,' said Ian.

I was reluctant to do the whole packing routine again, but we needed to rent the flat to someone. Luckily, while our elder son, Jamie, had been home again for Christmas – this time from Buenos Aires – he told us some friends of his needed somewhere to live. A lovely young couple with a baby decided to rent our place furnished and have our younger son, David, live there with them too.

Care of the flat sorted, I began buying canned food and dry goods like rice, a wide variety of beans and pulses. We needed first aid items and seasickness remedies, and to pack our cameras, snorkelling gear, wet weather clothing, deck shoes, and sun protection. We also had to think of things that we might need once we were on our own – things we might not be able to get in Darwin.

I wanted to take a laptop computer with us. In 2004, I still didn't know much about computers and ended up buying a secondhand one without a modem. I thought I really just wanted a typewriter/word processor and was a little annoyed when I realised that a modem could connect me to email.

April approached and the day of departure eventually arrived. Our things were already stowed on board *Libelle*. Despite looking ill from the strain of the previous few months of hectic preparations, Tim refused our offer to bring *Libelle* down from Pittwater to Sydney Harbour. He wanted to sail to Elizabeth Bay with two close friends and planned to take us on board there. He wanted to be farewelled by his family and friends from the bay where he had sailed VJs (Vaucluse Junior yachts designed for young teens) when he was a teenager. The departure point held huge significance for him because it was where he'd spent almost all his free time while he was growing up.

On the 6th April 2003, nearly a hundred people turned up at Double Bay for a picnic lunch and to wave *Libelle* goodbye. Tim had many friends and only a few of those people were friends of ours. David came with us so that he could take our car home again. I was excited about going to sea again. I'd missed sailing during the two years since we'd brought *Libelle* to NSW.

At the park, I met many of Tim's friends, his sister, who'd travelled up from Canberra, and also his ninety-year-old mother in her wheelchair. She was worried she would not be around to see her son when he returned from his circumnavigation.

That evening, we settled in on board. Naturally, it took us several weeks to adjust to each other and our new situation as crew. Ian found things easier than I did. He is easier going than I am and he'd been working on *Libelle* with Tim for months. Not so me. I was used to discussing situations with Ian and coming to an agreement, then acting. On this voyage, Tim was skipper yet, during those first weeks, he seemed reluctant to accept that role when he was so inexperienced compared with Ian and me. He, too, seemed unsure of himself in his new venture.

It was not the best chapter of my life. I was unable to adapt to the changed circumstances. *Libelle* was no longer my home, but Tim's. I was used to taking control of things, but there was no place for that now. The only control I had was over the cooking.

Before we left Brisbane, I told Tim I thought I should go home and leave Ian on board to continue crewing.

'I didn't know you were unhappy,' said Tim.

Some personal issues had arisen for me during the voyage and I needed time alone to deal with them. Tim talked me into staying. 'We can iron out any differences,' he said.

I wished I'd left. Later, it became too difficult to find transport back home. I continued on board, becoming more and more disgruntled. I had been told some years before by my superior at work that I lacked adaptability. I had been astonished. I'd thought I was easygoing. Apparently, I was quite wrong. I could be very difficult to get along with, especially in close quarters, and it was Ian who was easygoing.

Tim enjoying sunset, Grahams Creek, Curtis Island

I had anticipated some time for the snorkelling I had grown to love when we reached the tropics and the coral of the Great Barrier Reef.

We had seen areas of coral bleaching in 2000 and I feared that climate change was going to bleach more of the reef before I could get to see it. But Tim was keen to push on to Darwin because he hoped to arrive in time to join the Darwin to Ambon rally.

Libelle *anchored at Scawfel, Whitsundays*

Part of me was still enjoying the sailing. The days were mostly mild, the winds good and the seas sparkling. *Libelle* sailed beautifully. Southeast of Cairns, I persuaded Tim and Ian to stop at my favourite island, Dent. It had not been overrun with high-end tourism like Hamilton Island in the Whitsundays.

One day, well north of Cairns, we encountered a warship. There was no ensign and nothing to show its national allegiance, merely a huge figure '1' on its side. It was a small aircraft carrier carrying several helicopters. I was certainly intimidated by this vessel, steaming along in passages usually commanded by yachts and fishing boats and I suspected that the guys possibly were too, but they didn't admit it.

As the vessel was approaching us, our VHF radio broke into voice, 'This is *Warship One*. We are constricted by draft to deep water. Would the Commander of the vessel at position 'Y' please take evasive action – Sir.'

The voice was unmistakably American and despite the politeness of the words, the tone was icy. The fact that we were not the vessel he was addressing didn't matter. His voice sent a cold shiver up my spine to tingle my scalp. The grey monolith motored south past *Libelle* at about 20 knots.

We talked about what it could be doing in those waters. How many Australians knew about this? Why was the vessel so far away from Cape Townsend where the Australian and US navies usually cooperate in joint exercises?

Further on, we anchored in the lee of a small island, an anchorage we shared with a prawning boat. Another night, we turned into the Escape River to gain shelter for the night. There was a pearl farm there on Turtle Island and I would have loved to go ashore to look at how they farmed the pearls, but Tim was still in a hurry to reach Darwin, and we departed early the next morning. As we exited the river mouth, I noted the red soil showing in the low cliffs bordering the river mouth.

'That's bauxite,' Tim told me.

'I've always thought of bauxite as grey, like aluminium,' I said.

When we finally reached the tip of Cape York, Tim seemed happier to linger. At some stage, he had received an email saying he was too late to enter the 'race' to Ambon. We stopped in a bay just to the west of the cape and went ashore to explore.

Neither Tim nor I had been to Far North Queensland before but Ian had been there by small motorcycle in his early twenties. It seemed weird to be standing on the northernmost point of mainland Australia. In this isolated place, there was the usual signpost showing the distance to various places around the globe and a hazy glimpse of land to the north.

Jan stands at Cape York

There are several rocky islets extending north of the cape, indicating a previous time of lower sea levels. Torres Strait is very shallow and this area is part of what was once a land bridge to New Guinea. We could see no signs of human visitation except four-wheel drive tracks in the dirt. Over the next few decades, even more of that land bridge will become swamped as sea level continues to rise.

Our next port was Thursday Island, close by in the Torres Strait. I had first heard about the island from Ian early in our relationship. Ian had gone bush soon after his university graduation, riding a Honda 90cc motorcycle to Cape York and west to Katherine. From Cape York, he'd taken a ferry over to Thursday Island.

I imagined it as a remote island imbued with mystery. After all, it lay only a few miles from Papua New Guinea, the mysterious country to our north, where once savage black tribespeople decorated with feathers had met with modern life just over a generation ago. Now I was about to visit that tiny island with the strange name. I was excited. My adrenalin soared when we had a close call on approaching the island.

Ian wrote:

We were heading towards Thursday Island before a light breeze. The mainsail was to starboard, the yankee (see glossary) poled to port and the *Aries* was steering. Jan and Tim were in the cockpit but I was the one doing the sailing. I changed the course on the *Aries* a couple of clicks, stood on the cockpit seat and took a good look ahead, then went below.

I tidied a few items and then plotted the GPS position on the paper chart. The little pencil cross was almost touching a reef. I stared at it, felt the hair rise on the back of my neck, thought of checking my plot, rushed up to the cockpit, yanked the *Aries* chain off the tiller hook and thrust the tiller hard to starboard. I climbed onto the deck, hooked my foot over the tiller to keep *Libelle* turning to port, and looked ahead. Forty metres away there was a brown reef, about a metre below the surface, extending at least 100 metres either side of the course we had been on. As *Libelle* slowly turned and the reef became visible to Tim, he intoned, 'Oh, shit!'

I had known we were heading towards the reef and was intending to drop the pole, turn to port and broad reach past the reef at a safe distance. I had calculated the time to turn more than an hour ago, but had not realised that the tidal current under our tail had since increased because I had omitted to compare the log readout in the cockpit with the GPS down below. More generally, I had been overconfident in reef-strewn waters with notoriously strong and variable currents.

At Thursday Island itself, there is nowhere to anchor, but there is a small harbour at nearby Horn Island and a ferry runs between the two islands. We went across to the town and I found it reminded me of towns Ian and I had visited in the West Indies in the 1970s, dingy towns without windows, just steel storm shutters to protect the shops from both theft and hurricanes – a practical solution to the frequent battering they receive from the climate, but off-putting to someone used to decorated shop windows and smart paint as in more temperate climes.

The next day, we spent hours ashore on Horn Island, where *Libelle's* original owner had been stationed as a pilot during World War II. There was a fascinating museum showing relics of the pearling days and the air base from which Australia defended the country from a feared Japanese invasion. It was a very informative place, taking us back to a time before our birth, awakening our imagination of those earlier times.

I was reminded that Peter Hooks, the original owner of *Libelle*, had told us how he'd continued his love of flying developed during his time stationed at Horn Island, by becoming a glider pilot. *Libelle* ('dragonfly' in German) was the name of a famous European glider.

Back on board, the three of us discussed whether we should stay a day or two until the wind returned, or continue westward using the power of the 'iron topsail'. I would have loved to stay at Thursday and soak up its ambience. However, my thoughts ran more along the lines that Ian and I could come back here by ourselves one day and, reluctantly, Tim agreed to move on the next morning. I was longing to get off *Libelle* and have time alone with Ian. We found we preferred to sail with just the two of us together.

Once again, the day dawned with a sky faded by the strength of the tropical sun. The only clouds appeared low on the horizon in the late afternoon, where the setting sun painted them vivid red, then after orange, pink, at which time the silky sea surface turned into watery blood. Suddenly everything faded and darkness descended as rapidly as it can do only in equatorial realms. The stars emerged and the moon rose. We motored on across the Gulf of Carpentaria in the stillness of the hot, tropical night.

The water of the gulf was still and flat as if it were solid, yet *Libelle's* hull, driven by the churning propeller, sliced through it like a butcher's knife through flesh. While we were crossing the strait, the wind continued to absent itself. Day after day, we motored westward. The days were tropically, glaringly hot, so that much time was spent inside the cabin. Occasionally, Tim decided to stop to swim.

Days passed and still, no wind came. We were reminded of Samuel Taylor Coleridge's stanzas from *The Rime of the Ancient Mariner*...

> Day after day, day after day,
> We stuck, nor breath nor motion;
> As idle as a painted ship
> Upon a painted ocean.
>
> Water, water, everywhere,
> And all the boards did shrink;
> Water, water, everywhere,
> Nor any drop to drink.

Except that we were not stuck, but moving under the power of the droning diesel engine. And we did have water to drink, clean water from stainless steel tanks, thanks to Tim who had installed them. The trance-like crossing of the Gulf of Carpentaria eventually ended and the wind arrived to sail us into Gove harbour and Nhulunbuy. The latter is the small town which serves the Aboriginal community and the workers at the aluminium refinery on the Gove Peninsula in the north-eastern corner of Arnhem Land. There was room for a few more yachts in the tropical haven of Gove Harbour.

We anchored securely, launched the dinghy and all three of us went into the town. Tim yearned more for ice-cream than the fresh bread and cold drinks available there. After seeing the small extent of the town, we walked to the yacht club, where we relaxed in air-conditioned comfort and chatted with other yachties, some of whom were living on their boats. Others, like us, were just passing through.

As the late afternoon cooled, we walked along the dirt ribbon leading to the Aboriginal settlement. Reaching the entrance sign to the village, we turned back. I was intimidated by imaginings of scraggly dogs and unwelcoming Aboriginal people. Mainstream Australian culture has done a good job of instilling an 'us and them' mentality. Probably, if we had gone in and asked to talk to the elders, we would have been made welcome. But we were too unsure of the protocol.

That evening, after the sun left our part of the sky, I gazed across the anchorage in wonder at the nearby, noisy refinery where the night shift continued working the bauxite ore.

The sky was alight with the glow of artificial light outlining the enormous factory like a gigantic birthday cake.

Gove aluminium refinery

The next morning, we moved *Libelle* into the industrial area of the port to take on fuel. *Libelle's* tanks were low after all the motoring we'd done since Thursday Island, and very soon, we were ready to leave again. This time, the wind deigned to blow, albeit gently. We could make way slowly under sail and relished the gentle swish of water as it bisected at the bow and swirled out at the stern of the hull.

Not far north of Gove Peninsula, there is a string of long islands jutting well out into the ocean in a north-easterly direction. Instead of having to sail north-east to round the top of this ancient mountain range, we were able to pass through a tiny passage between the islands. The experience of sailing through that passage was very different to anything else Ian and I had ever done previously.

We were approaching a long grey row of rocks that extended across the water as far as the eye could see. According to the chart, the rocky split was quite narrow – only a couple of hundred metres wide – and deep enough to allow a small boat to pass through. However, it was also tidal, with the flow reaching up to eight knots. That was as fast as the diesel motor on *Libelle* could push the boat, so we obviously couldn't make way against the tide; neither could we be pushed through too fast or we'd have no steerage and might be thrown against the rocks. It was best to start into the passage very soon after the tide had peaked, and move in the same direction as the water. It was imperative, therefore, to arrive off the passage at the correct time. Thankfully, the wind was light because getting this right was a challenge for both Ian and Tim. We slowed down our approach to a crawl across the water,

At first, it was hard to see the slit through the grey rocks. We kept moving closer and closer until it seemed we would crunch *Libelle* hard against the shoreline. A few hundred metres from the entrance, the gap between the islands appeared.

I relaxed a little, realising how tense I'd become, but then I tensed again as we closed with the entrance. The passage seemed so narrow that I wondered if we could fit through. It seemed to me that Tim's confidence in his ability to handle *Libelle* was much increased by this time and he took the helm. Slowly, he edged his boat closer until we slid in between two high, dark walls. Less than half an hour later, *Libelle* successfully slipped out into the open sea on the other side.

Ian pulled the reefs out and reset the sails. I also unleashed my muscles and I suspect Tim, and perhaps Ian did so too. We were now in the Arafura Sea with Arnhem Land (part of the Northern Territory) to our south.

Days later, we approached another peninsula. Tim had decided he wanted to explore Port Essington, lying abandoned on the eastern shore. He had read about the history of this place and was keen to see the old town of Victoria for himself. He nosed the yacht south until the GPS showed the appropriate readings. When we became aware of clearings ashore and a landing place, we anchored. We could find no way of contacting the ranger from whom we were supposed to gain permission to land there.

The next morning, we went ashore to find out more about this old settlement, once a thriving port and then abandoned. We wandered through the ghost village, finding old brick kilns and, further along the well-kept dirt paths, we saw the destitute foundations of buildings: stores, garrison, officers' houses and more, including the bakers' ovens.

At one time, thousands of European settlers and soldiers had lived there. Now, the site was eerily home to no-one. Long ago, everyone had moved to the fledgling town of Darwin. Wikipedia reveals that the settlement struggled from the start that it was rebuilt after a cyclone and was finally abandoned in 1849.

Deserted settlement, Port Essington, Victoria NT

The entry reports that in 1848 a British scientist wrote that Port Essington was "most wretched, the climate the most unhealthy, the human beings the most uncomfortable and houses in a condition most decayed and rotten".

For me, this stopover was most unexpected but it remains special in my memory and I relish the wonder it brought me and the vividness of history. When there was no more to see, we returned to *Libelle* and made preparations to move off again. Darwin was now tantalisingly close.

After we motored around the northern point of the peninsula, the wind came in again, slowly strengthening up to 25kts as we sped along with the tide assisting us. To the north lay Melville and Bathurst Islands and not much further westward, we entered the very large shallow bight of Darwin where the tidal range peaks at about 8 m. The sea is so shallow at low tide that one has to anchor about a mile offshore so as not to go aground. It was late in the day, so we anchored well out from shore, ate dinner and retired to bed.

At last, we had reached our destination and I was already anticipating the next stage of our big adventure once we were ashore.

Tim needed to find somewhere convenient to moor *Libelle* while he made his preparations for sailing to Indonesia. There were three marinas, we found, all within walking distance for fit people. Darwin is a compact city, lying close to the shore and easy to walk around. Tim decided on a marina and made arrangements with the manager.

The tides are huge and we had to pass through a lock to enter the marina pond. Entering the lock reminded me of entering the rock slit hundreds of miles back at the eastern side of the Arafura Sea. I experienced the same surge of anxiety as we approached a sheer rock wall, only to find an entrance at the last minute. Tim was being directed on VHF radio by the dockmaster, who was watching our approach from his tower beside the lock gates, so I knew it was unlikely we'd hit the wall, but that didn't lessen my anxiety.

How fearful we humans can be of the unknown. The last time Ian and I had entered a lock was in the Panama Canal, back in 1977, when we came back into the Pacific on *Caprice* during our circumnavigation. I was nervous then too.

We motored through the gate. Tim put the engine out of gear and we waited a few minutes while the dockmaster turned on the water pump so that *Libelle* rose along with the incoming water and the gate in front of us opened. We moved into the marina, which was compact, with apartment buildings looking down on it from opposite sides and a few shops on the other.

The marina in Darwin

At our allocated berth, Ian and Tim tied up the yacht and I packed our things. We needed our own space again and so did Tim. Besides, his next crew member was arriving soon.

For a few days, we stayed on another boat in the same marina. Although we hadn't met them before, the owners of *Byamee* were also members of the CCCA. It was fortunate for us that Peter and Ruth were happy for us to be 'yacht sitters' while they flew to Sydney for a week.

During that week, we seldom saw Tim, even though we were parked close by in the same marina. He was busy with his preparations, and we with ours. Once we left the marina, we didn't see him again until he returned from around the world a couple of years later, but we kept all his newsy emails and made contact upon his successful arrival home.

A couple of years after Tim's voyage, he published a well-written and glossy book about his two-year return, amply illustrated with his wonderful photography. (*Sail me Round the World*, by Tim Lamble, 2012.) Tim's circumnavigation was a great achievement for someone who had never been on an ocean passage before setting out. But then, Tim is not a person to be easily intimidated by a practical challenge.

⚓

In Darwin, we soon found and bought a Toyota four-wheel drive Land Cruiser camper. It was the only one for sale, so we had little bargaining power. We moved into it at a camping ground while we completed the necessary work on our list of maintenance tasks before heading into the wilderness. It was bliss to be just the two of us again. I had my control back!

We grinned happily at each other as we drove south out of Darwin to drive to Katherine where we were joined in by best friends, Judy and Mike Handlinger, in their big Ford ute with a slide-on camper.

Together, we explored the extraordinary gorges at Mataranka and Katherine. I particularly remember the pool where we swam in the creek at Mataranka. Trees lined the banks and their overhanging leaves dappled the water surface, protecting us from the harsh tropical sunlight.

Judy and Jan cooling off in the Roper River, Mataranka

In the Katherine gorge we joined a river cruise, photographing the colourful rock cliffs, and later bushwalked along the top of those cliffs.

At Katherine Caravan Park, we saw a brolga. I hadn't seen one before and I was very impressed by its dance. Was this an omen that we needed a brolga in our lives?

Brolga in Katherine Caravan Park

From Katherine we drove on to the magnificent Kakadu National Park, followed by a few days in Litchfield National Park. All of these places we found wonderfully exotic, magical, and the wildlife something special to see.

After Judy and Mike departed to travel south again, Ian and I continued across the north into Western Australia. The land was hot and dry, its undulating countryside interspersed with near dry watercourses and scattered with boulders and occasional trees.

⚓

First stop in Western Australia: Lake Argyle. We went for a cruise on the lake, which I loved. The thought of drowned farms and villages beneath us fascinated me. I gazed over the side of the boat, occasionally seeing dead trees down there, but I was searching for buildings and I wondered how the people who had lived and worked there had reacted to being ordered to leave their homes and community. I am sure they would have protested, as would I.

Travelling through the Kimberley in the dry season was an experience which captured my soul. I have long wanted to go there by water too – especially to King George Sound.

One abiding experience occurred latish one afternoon when we drove to the top of a knoll, where we found an area for stopping overnight. There was a water tank and a toilet provided for campers. There was no way a bloke with a folding army shovel could have dug a toilet hole in that hard, rocky surface.

When we pulled up, we saw a four-wheel drive and caravan parked. Inevitably, in such isolated country, we began talking.

'Are you intending to stay the night here?' asked the woman.

'I think so,' I replied.

'I'm not sure if I want to,' she said.

'Why not? It looks a great spot to me. Look at the view,' I said, gazing out across the landscape for miles and miles.

'I get worried in such isolated places,' she replied.

I couldn't comprehend her attitude. For me and Ian, to be alone together, to commune with the land or the sea and the immensity of the sky is an intense privilege – one mostly denied to those who live all their lives in cities on this increasingly crowded planet. What danger did she envisage? Suddenly I thought maybe she was scared of us?

On our way down the west coast, we discovered that our knowledge of Western Australia had been limited. We hadn't heard of Ningaloo, the western barrier reef, where coral grows close to the shore, nor of the oldest and largest living fossils on earth, stromatolites, a little further south. Monkey Mia and its dolphins did feature on our radar though.

Ian beside Land Cruiser, westernmost point WA

During our trip around Australia, we hunted for another boat more suitable for our needs than *Libelle*. From Geraldton onwards, we looked in every boat yard and every marina. In particular, we were looking for an older design – a Brolga 33 or something very similar, perhaps an S&S 34. Our next boat had to be a proven ocean sailer with a heavy keel. She also had to be comfortable inside, with a fridge/freezer or room to install one. Other essentials included a double berth, sea berths close to the centre of gravity of the vessel, headroom in the saloon and comfortable seating. We wanted a diesel motor and preferably not gas for cooking.

As well as boat-hunting, we were keen to see the famous wildflowers of Western Australia in spring. They were wonderful and I took two large photo albums of photographs of those beautiful blooms. Even crossing the Nullabor, we saw wondrous wildflowers, though many of these needed to be viewed with a magnifier and photographed with a close-up lens.

Wildflowers in WA

In the south-west of the state, we camped one night in the Boranup Forest surrounded by extremely tall Karri trees. Again, we were the only people in the camping area. The rangers left firewood for campers beside the basic fireplace. It was a memorable night to be lulled to sleep beneath that high canopy with the breeze from the nearby Indian Ocean creating a gentle suspiration through the leaves.

A highlight of the Nullabor crossing was a detour down to The Eyre Bird Sanctuary on the coast of the Great Australian Bight. Once again, I was thankful we'd bought a four-wheel-drive. It would have been impossible to get ourselves to the sanctuary otherwise. The track was steep, winding and sandy.

As much as seeing the birds there, we were fascinated by a large shifting sand dune. During one half of the year, it moves about twelve metres in one direction when the prevailing wind pushes it that way; the other six months of the year, when the wind blows in the opposite direction, the contrary wind moves the sand dune back.

As we watched, the wind lifted a fine cloud of silvery sand into the air, showering it down the leeward side of the dune, on and on. Nature – it provides more awe-inspiring experiences than anything man-made.

Shifting sand dune, Eyre Bird Sanctuary, Nullabor

Before the eastern end of the Nullabor Highway, there are six or so lookouts over the Southern Ocean from which Southern Right Whales can be seen during their spring migration to their calving nurseries. We stopped at every lookout to gaze from the cliffs at the leviathans in the water below. Some of them were already accompanied by a calf.

Lately, I have signed petitions against the drilling for oil in this area. Sea-life is endangered enough in these days of rising sea temperatures, without whale nurseries like those in the Great Australian Bight being further endangered by outdated, twentieth-century industry.

In Adelaide, we stayed overnight with Ian's much older sister and brother-in-law. It was a happy time to catch up with his family members and meet his great-nephews and great-niece.

Ian's brother-in-law proudly showed us his sustainable energy efforts. He had installed a large array of solar panels on his house, and also tanks to collect rainwater for his vegetable garden.

When the new millennium opened, I had great hopes we'd see the end of fossil fuels for transport and electricity generation, the phasing out of fossil fuel extraction and the opening of the market for new battery technology, solar, wind and tidal energy. Sadly, this has taken decades more to happen than I expected. I thought that countries would join together to push back against the threat. Sadly, mankind seems frightened of change and slow to accept the challenge.

Ian and I discussed how yachts of the not so distant future would be powered. Some people were already powering their auxiliary boat engines with second-hand vegetable cooking oils instead of diesel and using batteries to store solar and wind generated power for use when the wind didn't blow. These technologies, we knew, would become ever more popular. Like us, our brother-in-law was disappointed that the government wasn't moving more quickly on climate change. In 2019 at the behest of the coal industry, it is still dragging its heels.

From Adelaide, we drove on the main route east, diverting down the Eyre Peninsula to see the sea lion sanctuary and sample the renowned Crystal Bay oysters. Another highlight was a personal tour to an Aboriginal archaeological site a few kilometres off the coast from Portland, where the local elder showed us the pre-European settlement remains. There had once been a village where only the stone foundations of the cottages remained and the remnants of an eel farm from which the villagers had traded smoked eels as far distant as present-day Northern Territory.

In Victoria, there were the once Twelve Apostles to see – nine huge standing rocks being eaten away by the sea. And before long, we were back in New South Wales, not having found the boat we were looking for. Would we find her in time to go cruising again?

3 Finding *Osprey A*

One day in April 2004, I happened to see an advertisement on the internet for a Brolga 33 located in Scarborough, Queensland. It had been posted only that day and I was the first to phone with an enquiry about her. I liked what I could see in the photographs.

We flew to Brisbane and hired a car to drive north to Scarborough. On our arrival at the marina we found the boat sales office, introduced ourselves and were escorted through the large marina to view the yacht. I stood on the walkway and checked out her name – *Osprey A*. That sounded okay to me, though I did wonder about the A on the end of her name. I later learned that she was Australian registered for voyaging overseas. Each registered vessel named *Osprey* had to be distinguished from any others, thus the 'A'.

Our first view of Osprey A at Scarborough Marina, Queensland

The broker led us on board and unlocked the boat. Standing inside the cabin, I knew at once we had found our boat. It was a similar feeling to the one I'd had so many years before when we first stepped inside *Caprice*. Was it something to do with both boats being built by Geoff Baker at Fibreglass Yachts in Sydney? There was a solidity to her that I liked immediately. I knew she could be our floating home.

'I'll leave you to inspect the boat,' the broker said. 'Come back up to the office when you are done.'

Alone, Ian and I looked at each other.

'This is it,' I said. 'This is our boat.'

Ian didn't want to be railroaded by me a second time. He felt that I had done that with the purchase of *Libelle* and he was wary of my doing the same again.

'Let's have a thorough inspection first,' he said. He went on deck and carefully started to examine the rigging and other fittings while I examined the interior stowage with equal care. *Osprey*, we learned, had been out of the water on a hardstand for at least twelve months while the owner had returned to Denmark to attend to family business. Ian was keen to see what deterioration had set in while the yacht was unused.

He searched for tags in the stainless steel rigging wire, wear on the lines, the adequacy of the bow roller, anchor winch and other fittings. Having completed his survey of the deck, he came below and removed the engine cover.

'Look at this, Jan,' he said. 'Someone has carved into structural timber to fit this engine into the available space. I'm not impressed with that.'

'Surely it's more important that the engine is properly mounted and that the prop shaft isn't bent?' I replied. I couldn't believe that such a solidly built boat would be compromised too much by this action.

Ian looked carefully at the other modifications of the cabin and realised that a pilot berth had been removed to create space for a book shelf and a cupboard which held a stereo player. The remaining berth had the batteries installed underneath. The base of the pilot berth had also been structural. The timber frame had been cut through and removed.

'Go out and walk on the foredeck,' Ian said.

I stood on the foredeck and felt the sponginess of it. I walked back and forth, and then I went back to see Ian.

'It's a bit of a trampoline out there, isn't it?' I said. 'What kind of core does it have?'

'I'm not sure. It might be balsa or it could be foam sandwich.'

'Which would be easier to fix?' I asked.

'Not sure. I think we should call it quits for now. Let's go back to the caravan park and we'll come back tomorrow.'

As we walked back towards the office, a fellow spoke to us. He must have seen us coming off *Osprey*.

'Watch out for boats owned by kitchen carpenters,' he said.

I was mystified at the time. Only later did we learn that the young Danish owner of *Osprey* was a carpenter employed to install kitchens.

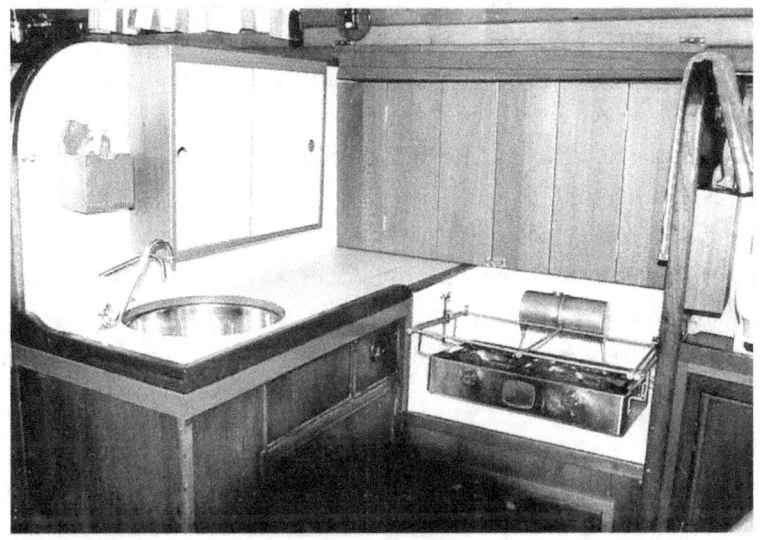

Osprey's *galley*

That night, we decided that the fibreglass was thick enough and the Brolga over-built sufficiently that we were willing to accept the alterations to structural timbers in those two places and that proved to be.

We did all the right things. We engaged a surveyor, a rigger and a diesel mechanic to inspect areas of their expertise whilst we looked over their shoulders. There were also phone calls to several patient CCCA club members who are familiar with Brolgas. We had her hauled out of the water to inspect the hull. Ian went up the mast to inspect fittings there, and apart from one broken strand in a shroud, we could find no obvious problems with the rig. There was a drawing of the unusual single spreader rig signed off by Joubert, the designer, and this led us to initially accept it as adequate, despite a very shallow angle on the lower shrouds and the chain plates of the lower shrouds not being tied in directly to the ring frame, as they are for the more usual double spreader rig. We later found that the lower shrouds had to be tensioned rather frighteningly tight to stop the middle of the mast sagging to leeward, meaning that the most highly stressed rigging wires had the least chain plate support. It was clear we would have to replace the rigging wire and maybe also alter the way the mast was supported. Ian made a note to replace the forestay too, because of the weight of the furler it carried.

Knowing that the deck would need extensive work to strengthen it, and that the engine possibly required expensive maintenance through being left unattended for over a year, we did some rough calculations and made an offer considerably below the asking price. Our realistic offer on *Osprey* was accepted with only a small adjustment. Was that a warning that the owner was aware of the lack of maintenance on his boat or was he just desperate to sell?

We moved on board, staying for a few days to sort out details of work to be done before flying back to Sydney. I was happy, but not in the hyper-manic way I had been with *Libelle*. We'd found our Brolga. We drove our car from Sydney back to Scarborough bringing tools. That's when the hard work began.

⚓

After Ian had replaced the two rigging wires using *Staylok* fittings, the rigger signed off that the rig was serviceable. We were surprised to find that the mast had no bearing material supporting it where it passed through the deck. Ian had plans to alter the rig before we went ocean cruising again.

We found that the too-flexible foredeck had been constructed of foam sandwich, not balsa. Our investigations also revealed little adhesion between the top of the foam and the bottom of the upper fibreglass skin, yet core samples showed the foam itself to be in good condition. We eventually found a time consuming but simple way to remedy this.

We had some questions about the auxiliary engine, a *Diecon* 28hp marinised *Kubota* diesel, which had been fitted two years earlier. We had also noticed that one winch was seized up, but that didn't seem like too much trouble. After all, the young owner had sailed to Vanuatu and New Caledonia, before leaving *Osprey* in storage for a year. A year didn't seem long. Things couldn't deteriorate too badly in that time, could they? Either they could, or the young man had simply neglected all mechanical maintenance. He is a carpenter, and it soon become apparent that if he couldn't solve a problem with woodwork, he either got assistance from tradesmen friends who were not sailors, or he didn't bother.

We discovered that every shackle on deck needed lubrication to free it up. All the winches needed servicing and only one was self-tailing. The motor had been installed with the control lever connected incorrectly to the gearbox so that *Osprey* moved astern when the control lever was moved forward.

Osprey *on the hardstand at Marmong Point Marina, 2008*

All the mechanical work fell to Ian. Three days with *Osprey* on the hardstand to replace the very worn feathering propeller and install a log and depth sounder became nine days. Closer inspection had revealed that the engine required new mounts. One was badly bent and another had a broken bolt held together with a *Nylock* nut. The intermediate bearing for the prop-shaft was too close to the gearbox. Ian removed it, installed a flexible coupling at the gearbox, and replaced the packing gland with a new floating seal.

After five days of crawling on his right side into the narrow space to access the back of the engine compartment, Ian began referring to it as 'going into purgatory!' I was very relieved he didn't give up.

Apart from acting as Ian's gopher, I scrubbed the slime off the hull, only to discover that the antifoul was so thin the hull required repainting. The hull had been painted only ten weeks before. I applied another two coats of antifoul. I also spent hours taking out excessively long zinc-plated screws and replacing them with bolts or appropriate length stainless steel screws and, later on, we built a proper front with closing doors for the lockers behind the back of the port settee.

Once back in the water the days continued to pass as, with each job we attempted, we found yet something else done wrongly, inadequately, or that, through neglect, was just not working. We thought after two months of hard labour that we were ready to go for a short shake-down sail.

'I'll service the *Aries* first. You'll be seasick after not being at sea for a year and I don't want to be hand steering all the time,' said Ian.

Five days later, the penetrating oil had finally done its work, and with the assistance of an extremely long lever, Ian managed to get the supports off. To remove the central pin, he took the *Aries* to a workshop with a press. The men there applied eight tonnes of force to remove the pin. With that out, Ian could dismantle the gear, service every part and reassemble it. Probably it had never been serviced before.

We had been working all daylight hours for eight weeks, with only a few hours off during that time. We were so keen to get *Osprey* ready to take north for a few weeks that we neglected our own exercise and relaxation. Consequently, by the time we were confident enough to set sail, Ian was exhausted. The Chronic Fatigue Syndrome he has suffered from for many years resurfaces when he is under prolonged stress.

We had planned to leave on Monday, 21st June 2004, in company with *X Tempore*, a 34' X boat newly owned by our friends and fellow CCCA members, Ann and Nick Reeve, whom we'd first got to know in 1977 when we were both crossing the Pacific. Not knowing that we were there, they had surprised us with their arrival at Scarborough.

Nick and Ann departed as planned and we followed the next day. There was little wind as we motored across the northern end of Moreton Bay but the sails were drawing well by the time we reached Bribie Island. Ian looked more relaxed and was enjoying the sailing. I smiled and congratulated myself on pushing to get away for a couple of weeks R & R before we were due to drive south to attend to business in Sydney in July.

Ian had been getting despondent about the boat and the amount of work we still needed to do and I thought we should get in some sailing to remind ourselves of why we were working so hard. Now Ian thanked me for pushing to get away. We were enjoying ourselves, just pleased to be at sea again in our own boat, to hear the ocean's song.

As we neared Mooloolaba, Ian decided that we should continue sailing to Double Island Point. The breeze was good and from a comfortable direction. The *Aries* was coping reasonably well, though not quite as well as we could have wished, and our medications were keeping seasickness at bay. I cooked dinner and we ate before dark. Ian retired below and, securely tethered by my harness, I settled down to enjoy my first night watch in over a year. The stars came out and a sickle moon appeared behind the mainsail, creating a pathway of silvery light across the water. It was a beautiful evening.

Almost imperceptibly, the wind increased, and the *Aries* began to have difficulty coping. The positioning of the steering lines and block in the cockpit were not ideal, but we'd hoped they would 'do' until we found time to fix them. Likewise, Ian had poled out the genoa, noting the deterioration of the aluminium where the pole was fastened to the end fitting.

'Another job to be attended to before too long,' he said, 'but it will probably do tonight.'

I was soon hand steering, then Ian got up so that together, we could tie two reefs in the mainsail. By the time we'd decreased the sail area, I was seasick. I went below to lie down, taking a small bucket with me. An hour and a half later I got up again, to take over the watch and let Ian have some more rest. The wind had continued to increase, and a couple of showers of rain and some dark clouds blocked out the moon and stars, diminishing the earlier pleasure of the sailing.

The sails were not properly balanced and we were too tired to be bothered with the effort of adjusting them. Every so often, the *Aries* just did not recover and I struggled to help the boat back onto course. I didn't know why the steering had become so difficult, except that the tiller now seemed to be rasping on the wooden floor boards in the cockpit. With tiredness and seasickness taking their toll, I didn't manage to correct the helm fast enough and *Osprey* gybed.

I called down to Ian, 'Should I pull in the mainsheet before gybing back onto course?'

Ian, believing that the homemade boom brake was doing its job, told me to just push the tiller over. With an almighty crash, the boom flew across as we returned to our correct course. The boom brake was totally ineffective.

The next time I started to nod off and the sail was caught aback, I decided to stay on the new course to take us closer in towards Double Island Point. I was cold, miserable with seasickness, and my nausea was increasing with the effort of concentrating on the compass. Inevitably, *Osprey* gybed again, and this time, the headsail was also caught aback. Then came a third crash and the spinnaker pole was hanging from the genoa sheet and dangling into the sea.

Ian leapt from his bunk and without stopping to don his wet weather gear or harness, rushed into the cockpit. He moved up the side deck and grasped hold of the spinnaker pole, lifting it over the rail and onto the deck. I took hold of the aft end and helped lift it aboard, noticing as I did so that the end of the pole was ragged where the fitting was missing. Ian freed the pole from the genoa sheet-rope and then crawled forward to secure it in its chocks. The release line for the pole tightened. Without thinking carefully, Ian gave it a tug. The line immediately slackened, and an instant later there was another crash. The end-fitting from the broken spinnaker pole, which weighed about a kilo, hit the deck not far from Ian's feet. It had fallen from its attachment ring two metres above on the mast. Ian rubbed his head and grimaced.

'That was a near thing,' he commented.

It was indeed, especially as he had broken our night-time rule of never venturing out of the cockpit without being tethered by a harness. What could have happened to him if he'd been concussed on the foredeck didn't bear thinking about.

About 1.00am, we were off Double Island Point and, in the lee of the point, we furled the genoa and dropped the mainsail into its lazy jacks. Under motor, we tried to make headway across the bay to anchor in the shelter of the great sandhills. Huge clouds of smoke billowed behind us as the motor struggled to take us forwards. Now what was wrong? We'd used the engine just twelve hours before and it had been going fine. Would nothing ever go right on this boat? It was 3.30am when we eventually dropped the anchor and 4.00am before we climbed into bed. Ian looked haggard, his face grey with fatigue.

'I can't take any more of this,' he said. 'There's just too much work to be done. I haven't got the resilience to cope with it. I think we should sell her and go back to the unit.'

'Please don't make any such big decisions when you're so tired,' I said, hugging him. 'Let's get some rest and recover a little first, then we can discuss things more rationally.'

Fortunately, we were sheltered from the wind the next day, so we stayed anchored in the bay, resting and reading. During the afternoon, *X Tempore*, Nick and Ann Reeve's new boat, came in from Mooloolaba. They too had found the conditions offshore boisterous, but unlike us had no gear failures. The following morning they went over the Wide Bay Bar and into the Great Sandy Straits. We stayed put.

Ian took off the tappet cover and checked the clearances, then adjusted them to within the specified tolerances. There was no noticeable improvement in the engine. He also removed an injector that appeared not to be working. He checked the spray, which seemed normal. However, the engine was greatly improved and Ian was mystified as to why.

'Just be glad it's going,' I said.

Two nights of sound sleep and an improved engine gave us the confidence to proceed over the Wide Bay Bar. We opted to give Tin Can Bay a miss and motored straight on up to Garrys Anchorage at Fraser Island. We tried using the electronic *Autohelm*, which had given the appearance of working when it had been tested earlier. Now it refused to keep the course. Hand steering it was, something we both dislike.

Fortunately, the distance was short, and once the anchor was down, we remained there for the next three nights. While there, we laid a bead of silicone around the outer edges of the windows which were leaking into the lockers below, resealing them. The window over the galley had to be sealed again later.

When we woke early the next morning, the tide was right, so we decided on the spur of the moment to motor south to Tin Can Bay. There, we went ashore to the chandlery and ordered a new electronic *Autohelm* ST 4000 steering system. We also did laundry and luxuriated under hot showers at the caravan park, then a day and a half later, collected the new *Autohelm* and installed it. The previous *Autohelm* fluxgate compass and the tiller driver both appeared to be functioning well, so we left those alone. All that needed to be done was to change over the control box. I realised Ian seemed to have decided to keep *Osprey* after all; otherwise, he wouldn't have bought the new *Autohelm*. There was no more talk of selling her.

The cockpit floorboards were now a pile of timber lashed to the foredeck. They had moved during the overnight sail and had been impeding the free movement of the tiller – the cause of our steering troubles. Ian would have thrown them out, but they were native Tasmanian sassafras and I knew we'd find a good use for them later. I stowed them in the quarter berth.

With water and fuel topped up and a new tiller pilot, we headed out of the anchorage. We did the 'set-up' circles for the autopilot away from the other boats and motored over to Inskip Point where we anchored. The next morning, in calm conditions, we recrossed the Wide Bay Bar and motored back to anchor in the lee of those sand dunes at Double Island Point. The weather was warm and the calm conditions continued that day.

We had a forecast for suitable northerly winds of 10-15 knots the following day. About 10.30am, a light north-westerly breeze started up, so we got under way and sailed out of the bay. When we became tired of hand steering, we turned on the new autopilot which worked fine.

Although we had wind as evening approached, Ian was tired and opted to go into Mooloolaba for the night. About 10.00pm, we motored up the river and found ourselves space to anchor. We slept soundly, then arose at 4.30am to leave on an almost high tide. Before dawn, we were on our way south again in calm conditions. Despite the forecast for 15-20 knots, the wind didn't return and we motored the forty nautical miles back to Scarborough.

To our relief, nothing else went wrong on the journey back to the marina. We had been gone for thirteen nights and although the time for R & R had been very limited, we both read a couple of books, something we hadn't done for over two months. We'd also enjoyed several walks on

Fraser Island, delighting in the peacefulness and the birdsong. We felt refreshed.

⚓

With *Osprey* in her marina berth again, we drove south to attend to business in Sydney – medical appointments and to arrange to rent out our home unit. While we were in Hornsby, Ian was assailed intermittently by bouts of fatigue, and its attendant brain fog. Late in August, we packed up our belongings and stored everything in the double garage again.

When we returned to *Osprey* after six weeks' absence, it felt to me as though we'd 'come home'. We brought with us the *Aries* we'd bought new for *Realitas* in 1983. While we were in Sydney, we sold two old winches we had in storage and bought three new self-tailing ones to replace both of the jib sheet winches and the main halyard winch on *Osprey*.

At that time, Scarborough was a port of entry, so there were many foreign yachts visiting after crossing the Pacific. Many Americans sailed this 'downwind' passage and then sold their boats in Australia. One couple we became acquainted with were shipping their yacht back to the USA but off-loading gear. They sold us three American flexible solar panels which we tied to the deck when we were in port and these gave us a welcome addition to the power supplied by the two solid photo voltaic panels on the rear arch. Another couple was selling all their charts of the Pacific Islands, and we purchased these too. They included charts of New Caledonia, Vanuatu and Fiji, all places we thought we might visit in the future. Even if we didn't go there, our sons might. Those charts were cheap and would be useful.

The second phase of upgrading *Osprey* began. First, we tackled the deck. Ian had drilled a fine hole and ascertained it was indeed made of foam sandwich. Now he calculated the best method of reinforcing the gap between the glass layers. He concluded it would be best to inject two-pot epoxy glue into drilled holes at small intervals.

In preparation, we sanded the paint off the gelcoat on the deck. Next, Ian drew lines from gunwale to gunwale at fifty millimetres apart. He crossed those with intersecting lines from bow to near the mast at the same distance apart, making the deck look like a checker-board. Where each pencil line crossed, he drilled a tiny hole into the deck. We mixed epoxy and, using a syringe, injected a squirt of glue into each hole. Each squirt of epoxy hardened into an upright support and also sealed the drill hole.

When we were finished, the two layers of the deck were once again held firmly equidistant and the deck no longer bounced when we walked on it. I thought this was a clever solution of Ian's.

I was in command of myself now, unlike with *Libelle* in Marmong Point Marina, when I was incapable of making a decision to do anything. I worked as hard as Ian to make our new home robust and seaworthy. Once the deck was satisfactorily fixed, I started renovating the finish with paint. Just before our purchase, the previous owner had started painting the cockpit grey.

'Don't paint any more of the cockpit, please,' I had told him. 'I want to paint the deck and cockpit pale blue and white, not grey.'

Where it was easy to do so, we removed fittings. Otherwise, I sanded right up to them, removing as much old paint as possible, but not interfering with the original gelcoat. I spent hours masking the items that were not to be painted and chose a one-pot polyurethane paint – blue for the tread areas, which I mixed to the shade I preferred, and white for the rest of the deck. Instead of the sand we'd found too harsh as anti-skid on *Caprice*, or sugar, which didn't work for me in the paint I'd selected, I found a commercial product which would leave our knees intact if we skidded on it.

I painted all the non-tread areas around the stanchions and the gunwales first, applying two coats of white, and waited until it was thoroughly dry before masking off the tread areas, which were to be pale blue. I sprinkled the anti-skid material over the first coat of wet paint, before applying a second coat with the anti-skid mixed into the paint before application.

To our surprise, it took several days for the paint to harden, even though it was surface dry. I had thought September in Queensland would provide the perfect conditions for boat painting. The days were dry, humidity relatively low and temperatures were in the high twenties or low thirties during the middle of the day.

Osprey's deck is repaired and repainted

Jan repainting Osprey's *cockpit*

While I was working on the deck, Ian removed the gear lever and reattached it correctly. When the paint had dried, he installed the new winches and greased the others before reinstallation. He also checked and traced all the electrical wiring, labelling it clearly. When all his modifications were complete he drew an electrical diagram. *Osprey* had been built in 1969, before the electronic age, and we needed to adapt the navigation area to suit modern times. Ian made a lockable door for the cupboard next to the nav. station, so we could store our small printer in there.

We bought and installed an inverter in the battery compartment under the port-side settee to power tools, the computer and GPS, as well as charge our mobile phone. Ian cut out a stronger cupboard front for the starboard lockers behind the settee and cut out locker doors from it. We wanted to use those lockers for boatswain's hardware, and needed to be able to lock everything down for sailing out at sea. My job was to apply coat after coat of *Everdure* epoxy to the plywood, waterproofing it. After that, I installed the locker-front, bolted in the entire framework and screwed on hinges and sliding bolts for all four doors.

Using small bolts and nuts, we installed straps to tie down the locker lids under the settees around the table. Another task was to remove the carpet from the interior sides of the forward cabin. It had been glued into place and was full of dust. Rather than try to rip it off the ply, Ian made duplicate plywood linings and, after I had epoxied them with *Everdure* and painted them white, he fitted the new pieces into place.

The previous owner had fitted a large multi-door cupboard for clothing along the starboard side of the double berth, effectively turning the berth into a single bed. I convinced Ian that we needed to cut off one section of that long cupboard and reinstall the part with just two doors for my clothing. (He used another cupboard for his clothes). We put it well to the aft end (head) of the bed, which was seven feet wide. We already had plenty of head space, but with half the cupboard gone, I now had hip and leg space towards the narrow foot of the bunk.

Another improvement to our berth was to varnish the semi-bulkhead just aft of the chain locker. There is no deck opening into the chain locker; the chain comes off the winch and enters straight down a hawse pipe to be stored in the bow under the deck. The semi-bulkhead, being open at the top, allowed mud and water from the chain to splash onto the foot of our bed when the anchor was being winched in, so I devised a kind of canvas curtain to attach over the open area from inside. This enabled us easy access to the anchor winch motor and also to unsnarl any snags in the anchor chain, but it stopped splashes when the fabric was in place.

As soon as the toilet displayed a problem with its pump, I insisted we install a *Lavac,* which, with its large pump, is usually problem-free. Ian also dismantled the stove and while he installed new wicks in the methylated spirit burners I thoroughly cleaned the rest of it.

An important chore I accomplished was to measure and buy new mattress foam for the big double berth and also for the port settee. When the covers and curtains were washed, *Osprey* was looking much more cared for and homely. We were gradually achieving the standard we wanted for our cruising boat. I had no regrets that we'd bought her and, by then, Ian seemed happy about her too.

In mid-October 2004, we were finally ready to take a refurbished *Osprey* home to Lake Macquarie in New South Wales. We checked the tide and weather forecasts for Saturday and all looked good for fair winds to leave Scarborough the next day. Ian raised the radar reflector, tied down the anchor at the bow and set up the *Aries* steering lines to the tiller. Those lines were now fitted at the correct angle. We stowed large items securely down below decks, including the deflated dinghy, and packed away the mainsail cover.

I paid the marina bill and at 10.30 on a clear Sunday morning we were ready to sail. People we had befriended waved us farewell. I started the motor while Ian undid the spring lines. The south-east breeze was holding us against the jetty, so he undid the bow line, and finally the stern line, before jumping on board. He moved the gear lever to reverse and slowly we backed out of the berth, turned *Osprey's* bow towards open water and motored away. As soon as we were clear of the marina, I steered while Ian raised the mainsail, leaving two reefs in, and unfurled about half of the genoa. As *Osprey* heeled, we realised I'd forgotten to shut the seacock for the bathroom basin. About five litres of salt water gushed across the floor before I noticed and rushed below to shut the valve.

'I'll have to make a checklist to remind us of all the things to do before we set sail.'

Ian nodded emphatically.

Our sailing that day was short. We made our way across the sand bars of the shallow bay to Tangalooma. This is a resort island where seventeen old ships have been deliberately scuttled to create a sheltered anchorage and a dive site. The ships are slowly becoming colonised by corals and other marine life.

One of the ships scuttled for a reef at Tangalooma

We pulled in between the ships and the leeward (western) side of the island where there is a sheltered anchorage. There had been very little movement in the marina, and we needed a little more time to acclimatise ourselves to motion before heading out to the open sea.

That short sail revealed a small leak in the forward hatch, and a slight amount of slippage where the *Aries* steering chain gripped the tiller. Ian applied more silicone sealant to the hatch lid and drilled and tapped a screw hole into the fitting where the chain attached to the tiller, making it very secure.

By Wednesday, we considered it time to move on and motored out of our anchorage. Ian miscalculated the tide and we bumped on the sand – fortunately without causing damage. As we proceeded, the wind failed to appear and we motored north around the island, where we found a treat. Two humpback whales – a mother and calf – were cavorting for a tourist boat from the resort. We too welcomed the display of leaping and tail slapping. The whales made up for the fact that the alternator malfunctioned.

The fluxgate compass Ian had installed low down inside the boat was proving inconsistent as well. Something magnetic was obviously interfering with it. After all the other dramas we'd had, these two things appeared minor. Ian made a temporary fix to the alternator and we continued on our way.

Clear of the island, there was enough wind to make sail, although the sea was still a little lumpy from the southerly front that had passed through during the weekend. Once we moved offshore, the water temperature rose from 21° to 23°C, indicating we had entered the south-flowing current. That East-Australian Current (EAC) can be of great benefit, adding a knot or even two to the boat speed over the ground.

Before GPS, we had to take sun sights with a sextant (measuring the angle between the sun and the horizon at noon and a few hours before or after), to find an approximate position, mark it on the chart and then measure the distance between the points on the chart. We then compared this to the speed shown on the ship's log, or to the navigator's guess at the speed of the boat through the water. With GPS, there was no calculation required. We knew our position at any moment.

For all our first full day of sailing, the wind was from the north-west which meant that, due to the proximity of the coast, there was no build-up of seas. *Osprey* rose and fell gently with the swells. In the late afternoon a squall hit us, heralding the arrival of another south-westerly change, which made the wind and the sea more aggressive. Ian lowered the sails when the squall approached and continued under motor now that the wind was on the nose.

Later that evening, when we were off the mouth of the Clarence River, a thunderstorm approached from the north-east. At the same time, a pod of dolphins chose to accompany us, their bodies lit up in the water by the combination of lightning flashes and the light of the gibbous moon – scary, but also a fabulous moment. Following the thunderstorm, the wind returned from the south-west and we made good progress south. By the early hours of Friday morning, we could see the lighthouse on North Solitary Island, just north of Coffs Harbour.

When the wind dropped out again after breakfast, we motored for three hours, before being able to make sail. The forecast was for damaging winds and more electrical storms that afternoon, and we noted that the barometer was down to 999 hPs – fairly low for the NSW coast. Despite the strong wind warning, the wind continued fitfully from the south-west and we frequently motored when the sails would not fill. When the water temperature dropped to a fraction under 19 degrees, we knew we had lost the south-going current.

Slowly the wind continued on its normal anti-clockwise route and a light south-easterly was forecast to arrive during the afternoon. Apart from more dolphins off Port Stephens, nothing exciting happened until we were crossing the Newcastle Bight, headed for Swansea. The wind picked up and we had a brisk, white-knuckle sail across to Moon Island, which lies outside the entrance to Lake Macquarie. In the early hours of Sunday morning, we crossed the Swansea Bar and picked up a mooring on the seaward side of the Swansea Bridge. *Osprey* had sailed from Tangalooma in three days and three hours – much faster and more comfortably than *Realitas* would have done. Sleep was now in order.

Ian asked by radio for the bridge to open at six o'clock in the morning. We motored through and tied up to another mooring, waiting for the tide to rise before we continued down the channel and across the drop-over into Lake Macquarie. The build-up of wind and waves marking the end of our homecoming passage was like the crescendo in the finale of a symphony. Both raised our pulse.

We didn't bother about the sails, just motored across the lake towards Coal Point and on into Kilaben Bay. We anchored in the corner of the bay we came to call home for the next couple of years. The far western corner of Kilaben Bay was where our younger son, David, had anchored his ex-navy ammunition barge, on which he was living while he renovated it. This was the first time we'd seen the barge since it had been in the Georges River and David had managed to arrange for all the rubbish to be lifted off. He'd installed an engine and propeller system and driven it from Sydney's south, where he'd bought it, to Lake Macquarie – quite a feat with a flat-bottomed vessel which travelled at only two knots in very calm conditions.

A work in progress: David's houseboat interior. From L to R: Dagmar Cortis, Jan Mitchell, David Mitchell

David on board his houseboat

We were intrigued to see the progress David had made towards making this old WWII ammunition lighter into a home for himself. I was very impressed with how much he had accomplished in cleaning up what had been a floating rubbish heap. David had made the vessel semi-habitable.

The roof wasn't yet rebuilt but a few days after our arrival, a delivery of building materials arrived at the water's edge. When David and Ian had loaded the timber on board, David asked Ian to help him take the barge to the "F" Jetty, where he removed the broken roof and replaced it with a plywood covering.

⚓

We spent that summer mostly exploring the lake and surrounding area, and making friends with some of the locals. Naturally, with a 35-year-old boat we still found items to fix, but in general, after all our hard work in Queensland, *Osprey* was now in good condition and we were enjoying our ownership of this solidly built sailing boat. She no longer surprised us with maintenance issues and Ian no longer exclaimed with despair that he wanted to sell her.

Sometimes, it being the festive season, we anchored with fellow yacht club members, friends and acquaintances. There was much talk about cruising and places people had been. New Year is always a great time for setting goals and making future plans.

'Let's sail back to Tasmania,' suggested Ian.

'Good idea. We could go around to Port Davey and the World Heritage areas in the south-west this time,' I added with enthusiasm.

Thus, the plan to visit Port Davey was hatched.

4 South to Tasmania

Late summer is the best time of year for cruising to Tasmania, so we planned to leave Lake Macquarie about the end of January. However, early in the New Year I was suffering from sneezing, a sore throat and had a mild fever, followed a couple of days later by coughing. I'd caught a cold, I thought. That 'cold' was hanging on and the coughing persisted all through January. We were both feeling very frustrated about the delay. In early February I felt slightly better and was hopeful that I was nearly over the illness.

'Let's get going,' I suggested. 'I should be fine soon.'

We set off on 6th February, but I wasn't fine. It was a fairly miserable trip down the coast for me, with both seasickness and coughing putting considerable stress on my abdomen. Seasickness is a very uncomfortable business. First there is a mild headache, then a queasiness in the belly. I'd lie down on the saloon settee and press the appropriate pressure point on my wrist in a vain attempt to avert the vomiting to follow.

'Bucket.' I'd squelch the word as the contents of my stomach rose in my throat.

Ian would look at me and put the small white bucket into my outstretched hand. The bucket always smelt of stale seawater and soon I'd heave into it. When I was sure I was safe for a few seconds, I'd hand the bucket out to Ian who washed it over the side before handing it back to me, its wet cord dangling onto my bunk. When my stomach was empty, I'd rinse my mouth and wash myself, changing into dry clothing, and tuck up on the bed to sleep.

Whenever the wind direction changed or we needed to tack the boat, the process would repeat itself. This would go on for a couple of days. Sometimes a bout of coughing would bring on the nausea and vomiting.

Perhaps I really am a little crazy, but I'm willing to put up with this for two days in order to enjoy the remainder of the voyage and have the ability to linger in out-of-the-way places where people normally don't go. I actually enjoy being at sea in good conditions, and find peace in watching the never-ending swells slide under our hull, one after another after the other, like a lullaby.

We didn't try to make the voyage in one leg as Ian was practically single-handing. Our first stop was at Jervis Bay and the next at Eden. I felt very relieved when we entered Jervis Bay and anchored in the north off Montague Point shortly after midnight. At dawn – far too early – we were hailed by the Navy. Ian went into the cockpit to talk to the fellows in their big inflatable dinghy.

'We're commencing firing practice in fifteen minutes and your vessel is in our firing range. I strongly suggest you move now,' said the officer. We did so, despite our tiredness. By the time we were re-anchored, we were too wide awake to go back to sleep.

Travelling south again the next day, we kept watch for the south-going current (water temperature 22.5°C) and then came in to the coast again to rest in port at Twofold Bay. We spent a couple of days there at Eden waiting for the right weather and sea conditions to cross to Tasmania. It is a very safe haven for boats waiting to venture into Bass Strait.

I have a fondness for this south coast town, its large bay and its charming museum. Ian and I have visited the area many times, and our family even spent a summer holiday there with Judy and Michael Handlinger and their kids during the early nineties (See *Crossings in Realitas*).

Eden is a fishing port, but also the one used by many visiting yachts. Most people in boats call in to the friendly port to replenish fresh supplies. Back then, each morning at about 9.00am, the Port Captain posted a printout of the latest forecast from the Bureau of Meteorology on the noticeboard near the main wharf.

We spent our two days in port topping up water and diesel (Ian) and groceries (Jan). A cool change came through on the twelfth, so we moved to the southern shore near the woodchip mill for shelter. Soon after the southerly had passed, the wind came round to the south-east, and we prepared to leave. We raised anchor at 11.00am the next day as the wind moved further east.

A few hours after setting off that Sunday morning, Ian spotted the first albatross. These majestic birds have a huge wing span – up to two-and-a-half metres for the black-browed albatross. The amazing thing about albatross is that they are experts at using the air pressure troughs between swells to maintain their lift; they seldom flap their wings except in take-off. I love watching their elegant forms soar over the swells and I feel distressed when I think of how many are being lost at sea to longline fishing hooks.

In 2001 *Australian Antarctic Magazine* (Autumn 2001) reported that 'in some populations [of albatross] current death rates are unsustainable. For example, at the Falkland Islands, in the south-west Atlantic, about 17,000 black-browed albatrosses are lost each year, most likely to longlines in some other nation's fisheries. That's about two birds for every hour of every day!' Research has also revealed the dwindling numbers of albatross returning to their breeding grounds in Antarctica after migrating across the Indian Ocean to South Africa. Even after flying shorter distances searching at sea for food, many fail to return to feed their chicks. Man's greed destroys so much of our planet!

During the crossing of Bass Strait, we experienced fickle winds as we had the entire trip, frequently having to resort to motoring. The motion was not good. The residual swell from the south-east was jerking the yacht about, and the wind was fitful. It made life uncomfortable and for me, seasickness returned briefly. Even Ian wasn't really happy leaning over the navigation table to write up the log and chart our position, though he rarely throws up.

The overcast sky cleared in the early hours of Monday, and the wind came in aft of the beam.

Dawn watch is my favourite. In the log, I reported that the sky was clear and starry and that we were experiencing a small amount of favourable current. Unless it is very cold or wet, I love being in the cockpit at night, watching the myriad stars, listening to the hiss of the water sliding past the hull and looking out for other vessels.

But by midday the barometer was dropping again and high, streaky clouds appeared, indicating that another front was coming, probably within twenty-four hours. In a small boat, one is always aware of the weather and watching the clouds for indications of a change. In another life, I would like to be a meteorologist. Even now that I can no longer sail, I still keep a close watch on the weather.

By late evening, we were already across the strait and starting to sail down the east coast of Tasmania. The wind had backed through north to the north-west and we were anticipating another south-westerly. In the early hours of the next morning, the front passed over us.

We were travelling at seven knots with the triple-reefed main being the only sail up. The bow was slicing through the water which curled up either side of the hull and rushed down its sides.

The wind didn't stay in the south for long. Within the hour, Ian was able to unfurl a scrap of the genoa, and by the time we'd finished breakfast the wind was from the south-east again, and dropping. The rapid changing of direction of the wind meant that there were several different wave trains and intersecting swells, making for a less than smooth passage. By the middle of the day, we were experiencing light northerlies and, to my horror, we had an invasion of flies.

'Hundreds of them!' Ian reported in the log.

Having been raised in New Zealand where flies are much less prevalent than in Australia, I have an abhorrence of them in my living space. I keep a fly swat in the boat and I dashed about the cabin swiping at any fly that dared to land. Ian regarded my antics with the fly swat somewhere between amusing and tiresome. I wasn't satisfied until the last of those sticky pests was dead or out of my cabin.

The wind dropped further to only eight knots from the north. It was impossible to make way under sail in those conditions, and we were rumbling along under motor again. By dusk, we were at anchor, tucked in near Eddystone Point lighthouse on the north-eastern coast of Tasmania. We had been motoring or motor-sailing for seven hours and it was a blessed relief to turn off the noisy engine and blow any lingering exhaust fumes out of the cabin before night.

The next southerly arrived as expected and late the next morning, as soon as the wind had swung to the south-south-east, we were ready to continue our voyage south. With plenty of wind that morning, we sailed close-hauled with well-reefed sails. Only a couple of hours later, the wind moved further east and before long the weather was repeating the pattern of the previous day.

We anchored in Skeleton Bay that evening, near St Helens, to await yet another overnight southerly change. We were making slow progress sailing south, but I was still not well and Ian was doing most of the sailing himself while I slept, so he needed the extra rest too.

In such changeable conditions, Ian kept regular radio checks with the Coastal Patrol stations, firstly with Tamar, then St. Helens, and as we continued south, Scamander Marine Rescue. The coast station reported a half-submerged vessel near the entrance to Wineglass Bay and we took note of the reported position, knowing we'd likely be stopping in the bay.

The next morning, it was still blowing from the south-west, but we set off anyway with the sails winched in tightly and I cooked breakfast underway. At 10.30 there was so little wind that we started the motor again. Three hours later, we had struggled through an area of adverse current, and with a north-easterly breeze we were able to sail again, the *Aries* steering for us. We didn't understand why the current should behave like this south of St. Helens, and I wonder now if it is a tidal effect. We had experienced the same current before in the same place.

Ian reported seeing 'a large brown-speckled albatross, many gannets and mutton birds (sooty shearwaters) and a couple of tiny storm petrels, their feet dangling in the wave tops.' Part of the fun of being at sea is watching the wildlife. Earlier, we'd seen some common dolphins – they love to frolic and race each other in the bow wave when *Osprey* is sailing more than five knots.

Currents bring food, so life is more plentiful. We both like watching birds swooping and diving for fish, or just sitting on the water, as seagulls often do in social groups. These days, the East Australian Current (the EAC) is extending further and further down the coast of Tasmania, bringing plenty of food with it. The water is warming, too. Our thermometer often showed water temperatures of 24°C!

The northerly wind continued to hold all afternoon, strengthening towards evening. We shortened sail twice to keep *Osprey* under control with the *Aries* as the boat powered downwind. In the fading light we were quickly closing with Wineglass Bay. Ian double-checked the GPS position of the half-submerged boat. My shoulders were already tensing as *Osprey* approached the semi-enclosed bay.

Just outside the bay, Ian started furling the genoa. Suddenly, part way up the stay, the sail ballooned out. We were moving at five knots and the wind was still increasing. My heart began racing but Ian, as usual, appeared to stay cool. He struggled with the sail as it billowed and collapsed repeatedly. The whole boat was shuddering. I disconnected the

Aries and firmly gripped the tiller, hand-steering into the bay as Ian attempted to control the sail in the increasing darkness. I flicked on the depth sounder, but my attention was on Ian and that sail. Then the depth alarm pinged into my attention. I glanced at the depth reading – two metres! Our keel was about to touch the sand.

'Going about,' I yelled to Ian. I swung the tiller over.

'What?' Ian yelled back.

'Beach!' I pointed over the starboard bow as we moved away.

The shore was barely visible in the gloom. How had we come so close? I paid more attention to my tasks as I brought *Osprey* up under the shelter of the cliffs at the northern end of the bay. There, Ian finally freed the tangled sail. But the genoa flew fully out. I steered away from those cliffs. Our boat sped toward the opposite end of the bay. How could the area seem so small? In the cockpit, I was doing my best to haul in the furling line and steer at the same time.

Once Ian had returned from the foredeck, he completed furling the sail. I steered back towards the protection of the cliffs and Ian dropped the anchor. I backed up under engine to ensure the anchor was well dug in. We'd had enough drama for one night; we didn't need to drag anchor when the wind changed direction yet again. I turned off the engine and breathed deeply. We had overcome another drama!

Having tidied up on deck, we retired to the warmth and light of the cabin – Ian to report our position to Scamander MR and write up the log, me to cook a late dinner. It was with great relief that we finally sat down to eat and relax before bed.

As usual after such dramas, the next morning presented a completely different face to the world. I started to doubt I'd ever had reason to be panicking the evening before.

Just as in the tourist pictures, the water was blue in reflection of the sky. The bay is a gem with a long horseshoe-shaped, white sand beach on its western shore, high cliffs to the north and lower ground at the southern end of the bay. All was calm.

As we departed, we could easily see the small half-sunken boat just two hundred metres from the rocky southern shore. Above the water, the end of a spar projected, a triangle of light-coloured material attached to it

– the remains of a sail perhaps? Thank goodness it wasn't our boat wrecked there.

Wineglass Bay from the shore (picture from internet tourist site)

I spotted a tiny penguin battling against the half-knot of northerly current as it swam. I watched it while we set up sails and self-steering for a light north-easterly wind, and then we edged south-west, finding smooth water south of Cape Somerset. Light conditions prevailed all day, so we steered using the electronic *Autohelm* rather than the wind-operated *Aries*. The quiet conditions calmed our nerves after the hectic episode the previous evening. Our voyage south was presenting particularly variable conditions.

We anchored for the night in Spring Cove. There wasn't far to go now. We planned to take the short-cut route through the Denison Canal, the preferred route across to Storm Bay and to Hobart. The Canal cuts across the Tasman Peninsula and saves many miles of sailing.

We'd had a good wind when we raised anchor mid-morning, but it didn't last. We mostly motored through the channel to Blackman Bay, where we anchored ready to traverse the canal the next morning. After being at sea for days, it is quite fascinating to look at the different styles of houses, outbuildings and vehicles along the shore. Blackman Bay is bordered by the small village of Denison with mostly older-looking cottages.

Denison Canal opens at 8.00am and, that morning, the tide was high enough for our keel to pass through. When we exited into deep water half an hour later, the early morning mist was just rising and there was a light spatter of raindrops. Even when there is no mist, there is little enough to see from the canal itself.

Looking back at the Denison Bridge

On the radio, Ian informed Hobart CP of our presence in Storm Bay. Just a few hours later when we had crossed the bay, I spotted the landmark rock, the 'the Iron Pot' (a reminder of Tasmania's whaling past).

The Iron Pot, Storm Bay

We entered the River Derwent and by mid-afternoon we had anchored in Sandy Bay, near the Royal Yacht Club of Tasmanian. Hobart at last! It had been an exceptionally slow trip for us – fourteen days from Lake Macquarie to Hobart – but I was not well, Ian also needed a lot of rest and the weather had been unusually variable.

Osprey *moored at Constitution Dock, Hobart*

Ian contacted Hobart harbour authorities and booked us a berth in Constitution Dock, where we tied up next afternoon. Constitution Dock is right in the heart of the city, handy to shops, supermarkets, Post Office and buses.

There is even a good chandlery within walking distance. If we were going to head around to the south-west, we had to re-provision. This is not too difficult with my system of a small fold-up trolley, a large backpack and lots of octopus straps to lash the bag to the trolley. I must admit, however, that I do push the weight limits on the trolley. I have frequently carted twenty kilos of groceries back to our boat.

Only three days later, having topped up our groceries and water, we left the dock, sailed into the D'Entrecasteaux Channel and over to Kettering, where it was more convenient to refuel at the jetty, rather than cart plastic fuel cans through the city streets.

The nights were cold in southern Tasmania, and my cough became worse in the early evening. That's when it dawned on me that I was exhibiting the symptoms of whooping cough and I remembered that there had been an epidemic of it in NSW during January. But I'd had a serious dose of whooping cough when I was a young baby – and nearly died from it. Didn't that provide immunity for life? Apparently not. My GP explained later that you can get whooping cough several times over a long lifetime. By the time I recovered, it was already the end of February – the normal six to seven week duration for that disease.

We took a berth at the Southaven Marina in Kettering for two nights. Several people we know live in the beautiful village and we enjoyed catching up with them: John and Dee Deegan, Dave Davey and Annik Anselem, Dave and Maree Hoyle, and John Hamilton and Jean Taylor – all ocean cruising folk like ourselves and members of the Coastal Cruising Club of Australia.

In the nineteen-eighties, John and Dee borrowed our charts of the world when they set off on their circumnavigation in *Innisfree II*. Dave and Maree Hoyle spent many hours with us discussing their plans before setting out to cruise the Pacific Rim in *Byjingo*; they sailed in the wake of John and Jean in *Safari*, who had also sailed to Alaska. Dave and Annik's boat, *Windclimber* is another Joubert design, a thirty-four foot Magpie. They had sailed to many places around the Coral Sea, the NSW coast, Lord Howe Island and Tasmania.

Once we had completed all our social activities in Kettering and refuelled, we were ready for our next expedition – a voyage to Port Davey and the World Heritage wilderness area of south-west Tasmania.

The D'Entrecasteaux Channel is relatively sheltered from the Southern Ocean by Bruny Island, and the channel is narrow enough – maybe two or three sea miles or about five kilometres across – that there is plenty to see on both sides of the water. All going well, one could sail a small yacht right across the south-eastern coast of Tassie in a couple of days or maybe three, but we were on holiday and not in a rush.

We sailed *Osprey* across the channel to Barnes Bay in North Bruny, where we visited the Swans, a couple we'd met in the Queensland marina at Scarborough where we'd bought *Osprey*. They too had just bought another yacht, *Swanhaven III*, which had been on the hardstand next to

Osprey. Yachties make friends quickly and now I can't remember the first names of that lovely hospitable couple. We stayed in their bed and breakfast guest quarters overnight, while *Osprey* was tied up to their pier.

Swanhaven III *at anchor while we used their jetty*

A seal had broken open the nearby salmon fish farm pen, letting out lots of salmon. All the locals alert one another whenever this happens and the Swans had erected a small net on the point of their land, where several salmon had snared themselves the previous evening.

We ate barbecued salmon steaks for dinner that evening, washed down with a local wine, and later watched through the lounge room windows as a quoll came to eat the dinner scraps. What luxury!

Quolls are extinct on mainland Australia and relatively rare in Tasmania, but a remnant population survives on Bruny Island. They are about the size of a domestic cat, nocturnal, and have light-coloured spots on their backs. We had caught a glimpse of a pair in the dark one evening at Cradle Mountain during the nineteen-eighties, when we camped for the night with our friends, the Handlingers. The quolls were looking for food scraps near the campsite. Like the endangered Tasmanian devil, quolls are carnivores and have very sharp teeth.

Members of our yacht club were anchored at the northern end of Barnes Bay. Their Westerly 33, *Kyeema,* was at anchor in the area known as the Duck Pond. We anchored close by, had a cuppa with Nellie and Romul and then moved on to Cygnet on the main island early the next morning. Ian had had enough of socialising.

The sky was overcast and there was just a smidgen of raindrops, but no wind, so we motored north to Cygnet, where we dropped anchor close to the town wharf. The anchorage was excellent, protected and with a muddy bottom. The depth is close to seven metres at low tide. We had no qualms about leaving *Osprey* while we wandered the town. We took the dinghy ashore at the small sailing club, where we were made welcome by a couple of members who happened to be there. Cygnet is a peaceful and idyllic small town, green with English gardens, and nestled between the river bank and the shelter of a low range of hills to the west. It was pleasant to wander through the quiet town with gardens which reminded me of villages in England.

Late on Tuesday, after topping up our water, we motored back down the river and into the channel again where there was wind behind us. It is always so peaceful when the sails are pulling and we turn the engine off. We can hear the swish of the water along the hull and the squawks of birds on the wind. In gentle conditions like that day, the motion of the boat soothes and lulls as the sails push us along.

The weather forecast at 4.00pm alerted us that a west-south-westerly change was due, so we sought shelter at Southport before dark. There was no village there except for a pub, although there might have been a general store at one time. However, the fences along the roadside were covered with blackberries, ripe for the picking, and I gorged myself, arriving back at the boat with my face besmeared in blackberry juice.

Strong westerlies kept us aboard at anchor for the next two days. We sat one each side of the cabin, feet up, reading and listening to CDs of classical music. Once in a while, one of us would get up to put the kettle on or to look outside. Side by side at the table, we ate our lunch and did a crossword puzzle.

On Friday morning, although the sky was almost completely clouded and the wind had died, we set off. Recherche Bay was not far – just around the next headland – but it is a large bay and we needed to navigate right in, to gain shelter and have access to Fords Green, the village at the mouth of the creek, which empties into the bay at its southern end.

As we proceeded, the sky cleared, leaving a scattering of large cumulous clouds. It being still early afternoon, we ate lunch and then we went ashore at the area of low ground near the creek. As we walked around, I could see that there had once been more of a village there than was now evident – maybe a shop or two, a pub – but now all that remained were a basic camping ground and a few small cottages. These appeared to be holiday shacks or fishermen's huts.

Outside the cleared area, there was just scrubby bush and no proper roads. A rudimentary track went to Southport, which linked up with the rest of southern Tasmania.

We noted a signed pathway along the southern shore and we walked that the next day. The pathway led to a small rock platform, on which stood a bronze sculpture of a southern right whale. The sculpture was commissioned in recognition of the part whaling played in Tasmania's history. In our copy of the local cruising guide for the D'Entrecasteaux Channel, there was a note that a jetty was under construction not far east of the sculpture. We could find no sign of it in March 2005. I wonder how big the storm was that wiped it out, I mused, or was it never built?

Whale sculpture at Recherche Bay

We stayed a couple of days, until the strong westerlies blew themselves out. After the front came through, the forecast was for northerlies for the weekend and we were hoping to reach Port Davey before the next front.

We set off early in light conditions. As we rounded the rocky shore at Fishers Point, we saw a local fisherman pulling lobster pots into his boat. 'Yumm, I wouldn't mind one of those to share with Ian for dinner.' Fresh lobster is delicious, unlike crustaceans that have been frozen.

I was excited because I'd heard much about the wonders of the south-west wilderness. Heritage listed, isolated, no phone reception. Sometimes, people were holed up there by bad weather for weeks. Wonderful! We were already clear of the shelter from South Bruny Island. This was the Southern Ocean and the sea was lumpy.

At South Cape, the wind came in right on the nose. It didn't look likely to back into the west, so Ian decided to make a slight detour to Maatsuyker Island. From there, we hoped we'd have a more favourable wind the following day.

Once our course and sails were set for the uninhabited rugged island, we checked out the sailing guide regarding the anchorage there. The eastern bay, which provides shelter, also grows large areas of bull kelp making anchoring difficult, we read.

'What's bull kelp?' I asked Ian.

'It has extra thick, large and long, leathery leaves which can get caught up in the prop,' Ian told me. 'It'll make it hard to find good holding for the anchor.'

About mid-afternoon, we pulled into the bay. I went up to the bow to look for an area to drop the anchor while Ian steered according to my hand signals. Eventually, I found a small patch clear of kelp and I went back to the cockpit. Ian handed me the tiller and he went forward to drop the pick. The bottom was rocky, so we had to make sure the anchor was well dug in.

When I turned off the engine, I became aware of the seal colony. The bull seals were loud and smelly. Their colony was right at the base of the path and ladder leading up the cliff face to the lighthouse. I took out the binoculars and gazed ashore. The seals were large, be-whiskered, fat and lumbering – so different on land from how sleekly they glide through the water while swimming and hunting for fish. We didn't feel like confronting the seals to go ashore for a walk up to the lighthouse. Bull seals don't take kindly to intruders on their territory. Their barking, yelping and occasionally almost conversational noises continued throughout the night. That wasn't the only disturbance. Ian had set the anchor alarm and it went off in the early hours when the wind dropped away. For the remainder of the night, the anchor chain grumbled against the rocks on the sea floor and troubled our sleep.

By the morning of Saturday, 13th March, the wind had veered a little and was very light. We set off again, our sails pulled in tightly and the engine on.

During such light conditions, we do all we can to stop the boat rolling, the sails from slatting, and the sail slides from chattering against the mast. Whilst a deep keel will provide leverage against the water, to stabilise the boat we need to pull the sails hard in towards the centreline. That way, the forces against the keel in the water are balanced by the forces of wind against the sails.

The breeze came in by 10.00am and we were able to switch off the engine and ease the sails – always a more comfortable situation. Within a few hours, we were within sight of South-West Cape. Once we rounded that grey rocky headland – a foreboding looking place under the grey sky – we tacked and eased sail further, making it a pleasant run up the coast.

By 4.00pm, Ian was becoming concerned we wouldn't get in before dark, so he unfurled more genoa and we motor-sailed the last hour or so. As we came closer to the entrance of Port Davey, I could see why he had been concerned. There were many rocky islets around the southern shore. I took out my camera and photographed the rapidly changing shoreline.

Just then a fish took our trolling line. Ian hauled in the line along with a fresh mackerel for dinner. I found a bucket, half filled it with seawater and Ian dropped the fish in. I would clean it later. Ian refuses to clean a fish, but is happy to eat it if I prepare and cook it.

I could never hook a fish for sport and throw it back with a torn and bloody mouth. Neither of us likes to kill a living creature, but when it represents food, I harden my feelings. I try to put the fish out of its pain as soon as possible by stunning it with a rubber mallet. I give a blow to its head before I cut through its spinal cord.

The sun was low in the sky when we dropped anchor in Payne Bay, a large protected arm of water to the north not far inside the port entrance. Elation – we had finally arrived in Port Davey, a place I'd dreamed of visiting for many years.

The rocky entrance to Port Davey and Payne Bay

5 The Tasmanian Wilderness

Our mood was as buoyant as the bright, sunny-skied Sunday morning. Ian pointed out the Davey River to me on the chart.

'Can you see how the Davey River enters at the head of this bay? Take the binoculars and see if you can make it out.' I looked through the binoculars and couldn't make out a river entrance.

'I want to dinghy over there and find it,' Ian said. 'What I really want to do is to find out how far up the river we can get.'

'That sounds good to me.'

We packed the dinghy with water, spare petrol for the outboard, jackets, hats, cameras, binoculars and lunch, put on our life jackets and set off. It took about half an hour to reach the head of the bay and to find the river mouth.

Another yacht had anchored in the bay during the evening. Just before we lost our view down the bay, we could see the people on the other yacht also starting off in their inflatable dinghy.

The river banks were low-lying and covered in a monotone of brownish reeds, with occasional islands of similar boring, flat grasses. There wasn't even any bird life to be seen. As we moved further up-river, the landscape began to change. It became rocky and hilly, so that we were passing through a gorge. Beneath our dinghy's flat bottom, we could see the rocks below, the water stained brownish with tannin from the foliage.

The Davey River proved much more interesting than we had expected. It was, in places, spectacularly beautiful. We frequently stopped our little outboard motor to take photographs. The other dinghy passed us during one such stop. We waved and said hello to the couple aboard. Their motor had more horsepower than ours.

The sky stayed clear and the temperature continued to climb. Instead of putting on more clothing as we'd anticipated, we were stripping off. The temperature eventually reached into the high twenties and, to cool off, we were dipping our shirts into the river and putting them on again.

'Why didn't we bring sunscreen with us?' I said. We both laughed at the absurdity of such heat in mid-March at latitude 43° south.

'You know they say that climate change will bring more extremes,' said Ian.

'But they aren't predicting the tipping point for climate change to arrive until next year.' I grinned. 'But that's only a few months away. Maybe it's here now.'

The idea that people could question man-made changes in the climate seemed ludicrous to us. We'd watched its approach for decades.

Reflections in the Davey River

By the middle of the day, we had motored as far up-river as possible. It was time to turn back. The current was now with us and we could float downstream with just the force of the water until we were closer to sea level again. The other couple was having a picnic on the bank. We stopped to chat for a short while before carrying on. In no time at all, it seemed, we were in the last gorge.

A bird of prey hovered overhead and we watched until it disappeared from sight. Ian tried to guess if it was a falcon or a hawk, but the bird was up high and it was difficult to tell, even when looking through the 7x50 boating binoculars. The sky was clouding over by the time we re-entered Payne Bay.

The other dinghy followed us across the water. We tied up and climbed aboard *Osprey* just as the first strong gust of a southerly change swept in. We watched the others struggling against the wind and choppy water. They managed to reach their yacht and, relieved that they were safely aboard, we disappeared down *Osprey's* hatch to the security of our own cabin. Within half an hour, the temperature had plummeted by over ten degrees and the rain came, lashing the windows.

I boiled the kettle and we settled down to a cuppa. About half an hour later, when the wind and rain from the front had abated, Ian went outside to lift the outboard onto the stern rail and tie up the dinghy more securely for the night. As I started to cook dinner I reflected on what an excellent day we'd had.

At lunchtime the next day we anchored in Bramble Cove, where we ventured ashore for a walk up the peaty slopes of Mount Milner. The ground was boggy underfoot and occasionally slippery, but we made it to the summit, only 186 metres above sea level. There were tiny wildflowers and scat, probably from wallabies, though we didn't actually see any animals.

On our return to the dinghy, several yellow-tailed cockatoos flew low overhead. They are dramatic, raucous birds and I had not seen this black variety before. Their tail feathers were pale yellow and they had a bright yellow patch behind their eyes.

Later, we looked into tiny Wombat Cove which was occupied by a forty-five foot yacht called *Taipan*. Eventually we anchored for the night at Schooner Cove on the southern shore of the Bathurst Channel.

Between rain showers next morning, Ian and I went ashore. Ian needed me to cut his hair, but we also wanted to collect some fresh water and some mussels off the rocks.

On our way back, we called at *Taipan*, which had changed anchorage, to invite the couple to *Osprey* during the afternoon. The owners, Chris and David, invited us on board for a cuppa right then and promised to visit us mid-afternoon.

On the way back to *Osprey* I said to Ian, 'I look in awe at such big yachts run just with a husband and wife. I'm glad we didn't persist with *Libelle*. *Osprey* is so much more manageable for us.'

After Chris and David had visited us, I imagined they probably spoke about how they couldn't manage with such a small living space as in *Osprey*.

The rain had eased by Wednesday, so we motored right up the Bathurst Channel, through Bathurst Bay and into the Melaleuca Inlet, which leads to a tiny research station and the airstrip. Similar to the lower reaches of the Davey River, the banks of the inlet appeared to be of boggy peat and covered in low-growing reeds.

The creek was fairly shallow and someone had marked the channel with sticks pushed into the mud – not exactly a reliable marking system. We moved upstream slowly. *Osprey's* keel is deep, almost two metres, so we weren't surprised that we touched bottom occasionally. Suddenly we were stuck – hard aground on a rock just on the edge of the marked channel. The tide was rising, but the inlet is so far from the sea that we weren't expecting much tidal influence, certainly not enough to lift us off the bottom. I was surprised that we'd hit the rock because most of the channel appeared to be mud.

Ian was more concerned about getting us back into deeper water again. He tied the dinghy athwartships on the port side and tied a rope to the end of the boom. He pulled it out over the dinghy as far as it would go and then Ian and I both climbed into the dinghy. We grasped the rope and put all of our weight onto the boom. That didn't work. Then Ian tied a full jerry can of water onto the end of the boom and swung it out towards the middle of the creek. At the same time, we both jumped up and down on the bow. This time, success! *Osprey* gradually moved and we motored back into deeper water.

We had motored cautiously on for a few hundred metres up the creek when we spotted a mast above the low tree line. Around the last corner, the water deepened into a small basin and we could see the place to moor. There were three pile berths against the bank with telegraph poles deep into the mud for tying up. The middle berth was occupied by *Dovetail* and we pulled in behind her.

John and Shirley invited us aboard and over a cuppa we told them about our grounding. This was the first time we'd met. They were an experienced sailing couple about the same age as us, with ocean cruising acquaintances in common. We soon agreed to meet ashore after lunch to go for a walk.

The *Dovetails* had been up Melaleuca Inlet several times before, so they led the way to the Tasmanian Parks and Wildlife hut where we could see posters and information about the endangered Orange Bellied Parrot. There was a small Parks and Wildlife research station there. Next, they showed us the pathway through the bushes to the airstrip. The area surrounding the landing was flat and wet and covered in small mounds of button grass. Button grass is well named – it looks like a large, green, covered button. The only way to walk across it was to step from button to button or walk along the narrow, muddy tracks between the mounds.

Osprey, Dovetail *and* Taipan *moored at Kings Jetty*

A small plane circled overhead and then came in to land. The pilot had brought two hefty blokes in from Hobart. After they had picked up their packs, they told us they were intending to hike for a few days through the wilderness and then walk south-east to Recherche Bay.

When we arrived back to *Osprey* we met up with Chris and David who'd followed us up the creek in *Taipan*. We arranged to meet up later on *Dovetail* for sundowners. Our conversation that evening covered the usual topics that concern cruising yachties: people we know in common, where we've travelled, mechanical and electronic problems and toilets – always toilets. Put two cruising yachties together and within the hour the conversation will turn to boat toilets. A story will usually be told of a landlubber aboard who blocked the toilet by trying to hammer his beer can down, or some other improbable tale.

Our night at Kings Jetty, as the locals call the pile berths, was the stillest and quietest you could imagine on a boat. When we awoke, rain was pouring down. We had planned to leave and were pleased that the rain stopped by mid-morning.

'If you leave first, we will follow,' suggested John. 'We can pull *Osprey* off if she becomes stuck again.'

And so it came to pass... We duly hit bottom, became stuck in the mud and shortly afterwards, *Dovetail* came into view. Shirley slowed their boat while John prepared a towing line and passed it to Ian, who fastened it to our bow cleat. In just a few minutes, *Osprey* was riding clear of the mud again and Ian tossed the tow line back to John.

We motored carefully out of the inlet and anchored at Claytons Corner. By that evening, there were six yachts at anchor! There was an abandoned house ashore, which was presumably once inhabited by a family called Clayton. The old house now belongs to the National Park. It was open and used as shelter by hikers and visitors to the area.

Nearby, a track started up the hill behind the anchorage and Ian and I climbed to the top of the hill, named Mount Beattie.

View from top of Mount Beattie

My New Zealand mountains are high, rugged and frequently snow-capped, even in summer. Ian no longer bites when I rib him about the habit Australians have of naming hills as mountains. Mountain or not, I enjoyed the climb. The slopes were covered in small shrubs and grasses, but there appeared little in the way of birds or animals. From the summit, we could gaze down the length of Bathurst Harbour, which shimmered blue-grey in the distance. Its shape is almost square, which seemed very odd to me. We could see a schooner out there, her three masts clearly visible. She was making her way slowly towards the anchorage.

When we descended from Mount Beattie, Taipan's David was busy filling water cans from the rainwater tank beside the old house. The tank was full from the recent rain, and we thought it would be a good idea to top up our tanks too.

We returned to *Osprey* and Ian poured the water from our spare jerry cans into the boat's water tank. It's a fiddly exercise. I tie a tiny bucket to the air intake, which happens to come up behind the stereo. Then I watch the clear hose leading from the filler on the side deck to the tank. When the water backs up in the hose, I know the tank is full and I can let Ian know to stop pouring. By that time, some water has pushed up the air inlet line and dribbled into the bucket.

Ian took the two empty jerry cans ashore to fill them – giving us thirty litres spare. With the tank topped up, we wouldn't need to be concerned about water for about ten days. *Osprey's* main tank holds only one hundred litres, which isn't enough to do laundry, shower or wash our hair. That makes it sound as though we are dirty on board. We can wash the dishes in sea water and also wash clothes in seawater, but we use enough fresh for a final rinse to take the salt out. We heat a kettle of fresh water for an all-over wash every day. It takes but a cupful of fresh water to clean our teeth and the rest is for cooking and drinking. It is really important not to become dehydrated on board – or anywhere else for that matter – and we make a point of drinking at least three litres each per day.

The schooner we had noticed earlier came in to anchor too. There was a pile of firewood stacked on her deck. John and Shirley told us that the owners were locals who also ran the tiny tin mine ashore at Melaleuca. I am not quite sure how that works, having a tin mine in a World Heritage wilderness.

After another peaceful night, we motored away from Bathurst Harbour. There was a secluded bay opposite the large hill of Mount Rugby, which lay to the north. It is the highest peak in the south-west of Tasmania. Ian wanted to climb the hill, so we anchored. The weather was glorious that morning and I decided to do some laundry now that we had extra water. Ian prepared for his climb and set off in the dinghy to the other shore, while I set about washing the clothes and then myself as well as my hair.

Dried and re-dressed, I resumed keeping a look out for Ian on the hillside. Even with the binoculars, he wasn't easy to find. Three or four hours later, he was back down, feeling pleased with himself.

'I climbed to the top quite quickly,' he told me, 'but the summit was flattish. When I tried to come down, I found there were several tracks. It took me three goes to find the same path I'd come up!'

The next morning, we moved on down Bathurst Channel. Another yacht was approaching.

'That yacht looks familiar,' said Ian. 'I think it's *Sotalia*.'

'I think you're right,' I answered.' That looks like John at the helm.'

It was *Sotalia*. The Deegans, whom we'd known for years, turned around and we anchored together in Wombat Cove. John and Dede had their Shih Tzu dog with them and John showed off his dog-training tricks. The dog was more of a show-off than John. He was a really smart little canine with big appealing eyes and soft fur.

Both yachts stayed overnight in Wombat Cove. The next morning, *Sotalia* continued up to Bathurst Harbour, while we moved in the opposite direction, stopping in Spain Bay on the southern side of the entrance to Port Davey and opposite Payne Bay.

We were keen to go ashore and walk across to the ocean beach we'd seen when we first entered the wilderness area from the open ocean. The walk was worth the effort. Most city people have little awareness of the exhilaration experienced when one walks the length of a beach where there are no footsteps but one's own. In a crowded world, it is a blessing to own a boat that can take you to such wild, isolated beauty.

After coming back to our dinghy, we noticed that another vessel had arrived in the bay. By the time we'd arrived back on *Osprey* yet another yacht, *Sotalia,* was motoring towards us.

'Hobart radio has been trying to contact you. They said it was urgent, so we came back to let you know,' said John.

Immediately, Ian and I both thought of Jamie.

'If your radio can't contact Hobart Radio, the people on *Secret Affair* should be able to.'

Secret Affair was the yacht anchored some distance from us in Spain Bay.

'Then I'd better go over there and call Hobart.'

Spain Bay looking inland

He climbed into John's dinghy and they motored over to the yacht from Auckland. When John dropped Ian back at *Osprey*, they had been gone for over an hour. I was desperate to know what it was all about.

'The message is that we need to call the Anthonys. Jamie and Laura are overdue arriving in Panama,' Ian told me.

Our elder son was so much enjoying his life as a circumnavigator he was reluctant to enter the Pacific. Jamie had grudgingly agreed with his Australian girlfriend to transit the Panama Canal in 2005 and sail home. Now he was backing out.

'How are we supposed to make a phone call from here?' I asked.

'It seems the Anthonys have called the Australian Maritime Authority and the British consulate in Panama. Laura's travelling on a British passport.'

'I didn't know she had dual citizenship.'

'Anyway,' said Ian, 'Hobart Radio has been in contact with the Tasmanian National Parks. Those guys carry satellite phones. They've arranged for the Parks' rangers to meet us at 5.00pm as they make their way back to their base for the night. We have to position ourselves in the Bathurst Channel to intercept them.'

'Okay. You'd better get the anchor up. I'll start the motor.'

Ian went out on deck. In the forecabin I pulled free the long cable for the anchor-winch foot-control, which I passed out through the forehatch, then turned on the electricity to the electric anchor winch. I opened the engine cover to turn on the engine water cooling valve and closed off the drain from the exhaust-box before starting the engine. The anchor up, I began steering *Osprey* up the Bathurst Channel to our rendezvous with the park rangers.

It seemed like an hour but was only half that time before the inflatable dinghy with the two rangers came into view. They showed us how to work the satellite phone and Ian called the Anthony's home number. Soon he was speaking to Eden, Laura's dad.

It is always hard to correctly construe the import of a conversation when you can hear only one side of it. It seemed Laura's parents thought that *Possibilities* was quite late in arriving at Panama. Ian reassured them.

'It's very difficult to give an accurate time of arrival when you are sailing a very small yacht,' I heard him explain. 'The weather might have held them up. There might be too much wind or not enough. Jamie's outboard motor might have failed – again – or he could have run out of fuel. It's premature to worry. Jamie is very casual about keeping to a timetable.'

All that fuss to contact us! We didn't feel worried about our elder son. He's a very competent sailor. Besides, what could we do from half a world away? But Laura's parents weren't used to the loose timetable of a cruising yacht. They wanted certainty. It was understandable that they were worried about their daughter. Laura was to be bridesmaid for her sister and she needed to be home for dress fittings and other preparations, but we weren't told that at the time.

Ian returned the satellite phone to the rangers and thanked them. They headed off back to Melaleuca and we returned to Spain Bay. While I cooked dinner, Ian returned to *Secret Affair* to ask for a weather forecast.

We planned to head north-west the next morning to Macquarie Harbour if the weather still seemed favourable. Macquarie Harbour was the place the most intractable convicts were taken when Tasmania was a penal colony of Great Britain. There was no hope of escape from the hellhole of Sarah Island.

Until we reached Pilot Bay just outside Macquarie Harbour, there was no real shelter on the coast. We departed Port Davey at daybreak, all sail up, and motor-sailed hard on the wind into a very light northerly.

The conditions were mild all day and we alternated between sailing and motor-sailing, passing the two prominent landmarks on the coast between Port Davey and Cape Sorell. We passed by Low Rocky Point with its lighthouse early in the afternoon and, about 5.00pm in the evening, we were abeam Cape Hibbs.

Ian spoke to Mersey Radio at 8.30pm, when he gave an ETA for Pilot Bay of 1.00am very early the next morning. There was a three-quarter moon, which made the light airs and tricky sailing conditions just a little easier. At one point, when the wind came astern, we had to pole out the genoa – not an easy task in darkness.

Ian also used several waypoints on the GPS to ensure we were navigating safely into Pilot Bay and to clear the rocks that helped provide shelter from the ocean swell. We were both very tired by midnight and thankful of the moonlight shining on the rocks so that we were able to anchor a safe distance from them.

The Gordon and Franklin Rivers, south-west Tasmania, pristine huon pine forests: these names conjured up memories for me of the early 1980s. It was a time of protests, signing petitions, of car stickers reading 'Save our Forests' and 'No Dams'.

Then Bob Hawke, Prime Minister of Australia, acted. He led the Federal Government in an action to overturn the decision of the Tasmanian Government and its Hydro-Electric Corporation which had already dammed the upper Gordon and now wanted to dam the lower Gordon below its confluence with the Franklin. This act saved the two wild and remote rivers and enabled the whole of the wilderness south-western area of Tasmania to become World Heritage listed.

In 2007, *Cruising Helmsman* published my article about sailing into Macquarie Harbour and up the Gordon River.

Destination: The Gordon River
South-Western Tasmania

When Ian suggested that we sail to Tasmania in early 2005, I agreed on the condition that the trip should include the wilderness area of south-west Tasmania. Ian decided that if we could get to Port Davey, we might as well continue on to Macquarie Harbour, then up the west coast and around the top. That is, he planned to circumnavigate Tassie.

'Okay! You're on,' I agreed.

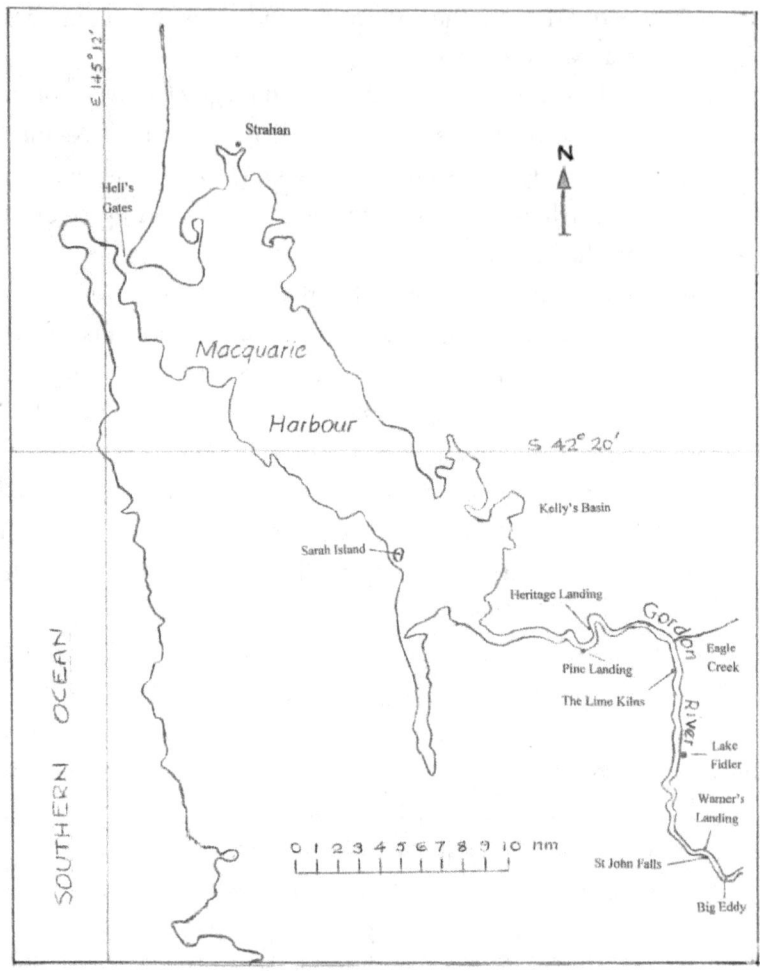

Fig, 2 Map of Macquarie Harbour

To visit Macquarie Harbour meant we also had access to those famous rivers, the Gordon and the Franklin. Autumn is the best time to make such as voyage.

An overnight sail from Port Davey took us up the west coast, to arrive before dark at Pilot Bay, just outside Macquarie Harbour and its entrance, Hell's Gates.

The entrance is narrow, with rocks either side. Off to the north is a long, low sand dune, possibly the entrance at one time. In more recent times, a seawall has been constructed, which helps protect Pilot Bay but reinforces the notion of going through gates. In strong westerly weather, with a heavy swell running, it is very dangerous.

Hell's Gates: entrance to Macquarie Harbour

Once inside the lagoon, the north-western end of the harbour is shallow, entailing the negotiation of a long marked passage which guides vessels around to the only town, Strahan. The Franklin River joins the Gordon River about 25 nm upriver from the Gordons' mouth, which flows into the south-eastern corner of Macquarie Harbour.

One story is that, during the early nineteenth century, second offence prisoners, who were sentenced to hard labour on Sarah Island at the southern end of the harbour, named this entrance as 'Hell's Gates', because life in the penal settlement was so brutal that many preferred death instead.

Strahan was originally built as a logging and mining town. It was particularly noted for its huon pine. Now, its chief *raison d'etre* is as a tourist town and gateway to the wilderness of the Gordon and Franklin rivers. While there, we needed to find local knowledge about the area. Trevor Norton and his wife Megs, who for several years have run a charter business in Strahan, were a fount of local knowledge of the kind we required. Trevor is the yachties' friend in Strahan. He has produced a very useful chart of the river.

We set off on Easter Sunday morning to travel the length of Macquarie Harbour. Sarah Island lies near the mouth of the River Gordon. On a mild sunny Sunday, it seemed benign and we had difficulty envisaging the hellhole it once was.

We walked all over the island, looking at the remains of the penal settlement. To assist our imagination, Ian lay down in the ruined foundations of an isolation cell no longer than his length and so narrow that he couldn't spread his arms.

Ian checks out a cell on Sarah Island

Late in the afternoon, we sailed across to Farm Cove where the settlement had tried, with mixed success, to grow its fresh food supplies. We anchored in a delightfully isolated bay, the silence broken only by a few bird calls, and slept well anticipating the trip into the Gordon River the next morning. However, the next morning, we awoke to light rain from cloud which enveloped us. The rain grew heavier and the visibility disappeared. There was no indication that things might improve, so we stayed at anchor. It was easier to imagine the miserable life of the convicts in this kind of weather.

Tuesday dawned fine, so this time we did enter the Gordon. The river is large and deep, navigable for twenty nautical miles upstream. The large tourist catamarans travel only a few miles to Heritage Landing. There is an illustrated boardwalk at this landing, providing an excellent introduction to the trees of the rainforest and its inhabitants.

Ian and I chose to keep going upriver while it was fine. We marvelled all day at the variety of greens, the marble cliffs, the isolation, the wonder of the river itself and, at mid-afternoon, we tied up at Warners Landing above which it is not permitted to take a keel boat.

Ashore, we found the remains of a Hydro Electricity Commission exploration site from 25 years before. The damage to the landscape was profound, but the forest was coming back.

Osprey *at Warners Landing on the Gordon River*

We decided to make the dinghy excursion up to the Franklin River junction the following day. On the basis of our experience in the Davey River, we decided to take lunch, extra water, sunscreen, a torch and compass (in case we found the aboriginal caves on the Franklin), walking shoes as well as sea boots and light wet-weather gear. There was also spare fuel and Ian threw in eight metres of rope. With all this gear, the two meters long inflatable Avon was very crowded.

We set off an hour and a half after sunrise when it was still quite cold on the shaded side of the river. The pilots from the float planes landing at St John Falls adjacent to Warners Landing had told us to stay on the outside of the curve, so they could see us clearly when they prepared to bring their planes down.

We had a 3.3 Mercury outboard, which took us over the ground at only about two and a half knots against the current of one and a half knots. The noise of the outboard shut out the sounds of any birds as well as the gurgling and splash of small streams and waterfalls coming down the steep sides of the river. The cliffs are densely clad, though where an old tree has fallen and gouged the surface soil it is easy to see that there is only a very thin covering over the rocks.

There were several dominant shades of green – the more muted green of the myrtle and myrtle-beech which shimmered when backlit by the sun, the blue-green of the eucalypts high on the ridge tops, the brighter green of the huon and other pines and a blue-grey from the occasional wattle. Leatherwood was still in flower on the lower slopes, its large simple four-petalled white flowers quite distinctive and aromatic. Festoons of fawn-coloured mistletoe hung from many living trees and fallen tree trunks bore a covering of thick green moss and lichen.

The most common birds we saw were ducks, which took flight at the sound of the outboard. There were also some cormorants and, very occasionally up high, we spotted a sea eagle with its snowy white breast.

We motored upstream for over an hour before we approached Big Eddy. Trevor Norton had warned Ian that we might have difficulty getting past this set of rapids with only a small outboard. We nosed about, looking for the best approach, but the current spat us out each time. Eventually, we managed to stop at the port bank and catch hold of some overhanging branches.

We took everything out of the dinghy, moving it about three metres at a time, up and over branches, across slippery rocks, sliding and squishing as we went. Nylon over-pants saved our trousers from the damp and mud as we slid over logs and down rocks. It was soon obvious from broken timber and squashed moss that we were not the first to cross upstream of Big Eddy this way. When the gear was about half-way, we went back for the dinghy.

The extra line tied to the painter was barely long enough, but we managed by tipping the empty dinghy on its side, and between the two of us – Ian lifting and pushing and me pulling – managed to bring it up to where we'd portaged the rest of our belongings. Then we went through the process all over again, moving everything another fifteen metres.

We continued motoring up a river that was becoming narrower, with more rocks above water and more current swirls. As we rounded the next bend, we saw that there was a second line of rapids ahead. I'd barely recovered from the hard work of getting around Big Eddy!

We tried one side, but couldn't hold our own against the force of the current, so we moved to the other side of the river. Again, there was no way of using the shore, but a huge log parallel to the shore gave us something to grab hold of. Using a small back eddy beside the log, Ian hoped to get enough power to take us over a curl in the current from which we could dodge sideways into a calmer spot. The risk of overturning seemed fairly high. Unfortunately, the outboard started to falter.

Changing the spark plug made no difference. Ian checked the fuel lines and found that the fuel flow was fine. It seemed the motor was just not going to cooperate. After many pulls on the starter cord, the motor would fire, but it died when load was applied. Sadly, we had not made it to the Franklin.

We started floating downstream, using the oars to keep in the faster flowing stream. Now we were in the silence of nature, drifting down this majestic river. We didn't we see a platypus although others have made sightings of these elusive creatures along the river, but this minor disappointment did not really diminish the joy of the natural world around us. Then, in a few seconds, we slid past Big Eddy, which had taken us so long to negotiate going upstream.

Over an hour later, we approached Sir John Falls landing, where the rafters who'd passed us earlier were drying out and sorting their gear. They'd been out rafting for nine days, having hiked overland to the upper reaches of the Franklin. If we had managed to get past the second rapids, they told us, there were no more rapids further upstream before reaching the junction with the Franklin. Just the confluence of the two rivers makes for some strong eddies, they told us.

Trevor was due to bring his sixty-foot yacht, *Storm Breaker*, up the river later that night to pick up the rafting party and take them back down to Strahan, leaving at first light the next morning.

Just downstream of the landing are a small beach and a hut erected by the HEC, but later taken over by the National Parks for walkers and kayakers. A party of eight Tasmanians with seven kayaks was camped for the night in the Parks hut on the opposite bank from Warner's Landing. They had come as far as Heritage Landing with one of the big catamarans and had paddled upriver that afternoon. In the morning, they were intending to try their skills over the rapids.

The next morning, the river was steaming with mist. We watched *Storm Breaker* depart with the rafters, and then soon afterwards the kayakers set off upstream.

Storm Breaker *returns with rafters from Sir John Falls*

Ian spent the morning trying to find out what was wrong with the outboard, but finally came to the conclusion that it couldn't be revived. He stowed it in the bottom of a cockpit locker.

We walked to the top of St John Falls in the afternoon. At the end of the bridge by the falls, we climbed over the bars, dodged around a tree trunk and started down the now 'disused' track that Trevor had told us about. A short distance along, a sign and a marker ribbon indicated that the falls were 'up'. The steep track was difficult in one or two places because there were no handholds, but mostly there were branches and roots to use.

After we'd been climbing for about twenty minutes, the track levelled out. We were glad of the pink marker ribbons, eventually finding a path to a mossy space above the falls. It was a magical place and well worth the effort to find it. We sat there amid the smells typical of damp, cool foliage for about twenty minutes in meditative silence before turning back. The return walk took almost as long as climbing up. In a couple of places, I slid down on my backside rather than risk falling.

After two nights and three days in all at Warner's Landing, we motored downstream only a few miles, then anchored near the three sulphide lakes. Lake Fidler is the closest to the riverbank of these unusual meromictic* lakes. We were keen to know more about them, so later I went to the Parks office to ask for information.

'You need to put an information board beside Lake Fidler.'

The man seemed astonished. 'Hardly anybody visits there,' he explained.

Maybe the Parks people would like to keep it that way, as the meromictic system is apparently very fragile. He explained the system to me.

Lake Fidler: a fragile meromictic lake

*Meromictic lake waters normally circulate their oxygen levels and that of other chemicals by over-turning once or twice per year. Normally, temperate climate temperature changes initiate the turn-over of oxygenated and oxygen-depleted gradients of water. Where the salinity gradient is sufficient to prevent a turn-over of the water, the lake becomes meromictic. Rare microscopic creatures live in this strange environment. In the case of the Gordon River meromictic lakes, the unusual situation of a deep sea water wedge intrudes under the fresh water layers, providing a salinity gradient strong enough to prevent turnovers. These lakes are particularly deoxygenated. As such, it is not surprising that the Sulphide Lake and Lake Fidler both appeared to attract very little wild life. The third meromictic lake in the area is Lake Morrison, which we didn't visit because it lies further from the river bank. The rarity of meromictic lakes was a strong reason for the World Heritage Listing. *(Source: Information from the Parks Office in Strahan.)*

Moving downstream we also visited Eagle Creek, where a bush walking track to the Franklin River starts at a campsite.

Early morning mist on the Gordon River near Eagle Creek

We walked up the track for half an hour or so. The under-storey of the forest was more open than further upstream and the thick moss lining the trunks of the trees gave a pleasant softening effect to the rainforest. The track eventually crossed to the other side of Eagle Creek, but the tree trunk which had formerly been the bridge had rotted away. Although several other trunks lay across the creek, none appeared to provide a safe crossing. Reluctantly, we concluded that we were not going to reach the Franklin this trip.

We stayed at that anchorage overnight, waking in the morning to a mystical landscape emerging in the mist. From Eagle Creek, we motored further downstream to our next stop at the old lime kilns.

We found the way up the cliff and there stood the kiln which had been restored a hundred years ago to make lime for the brick works at Pillager, the now abandoned town at Kelly Basin. We found remains which could have been at least two other kilns. The mosses and lichens were even thicker at this site than at Eagle Creek. There were many bricks scattered about, many half-buried in the bright-green moss. Some of the bricks were convict made, some more modern, and probably brought from the brick kilns at Pillager.

We drifted further downstream in the warm afternoon sunshine to Heritage Landing, anchoring in 12 m opposite the landing. Yachts are permitted to tie up there overnight, but not during the day because the commercial tourist vessels use it. Heritage Landing was very interesting and we wished we'd stopped on our way up.

The National Parks people have installed an excellent boardwalk several hundred metres long, with numerous information boards. Many of the native trees were identified, with details of how they were used by Aborigines and later by European pioneers. Insects and other local wildlife were also pictured and described.

Our final stop in the river was Pine Landing, but it bore little interest for us. It seemed to be predominantly a Parks rest camp for walkers and kayakers. With the sun low in the sky, we said goodbye to the kayakers there and set off to cover the last mile or so of the river. After motoring for another hour in the dark, we came back to Kelly Basin, where we dropped anchor.

How to get to the Gordon River

There are three ways:

1. By road to the tourist town of Strahan, then by tourist catamaran up the river or one can fly up to St John's Falls.

2. By hiking into the wilderness upper reaches of the Franklin River, and rafting down to the confluence with the Gordon, then on to St John's Falls.

3. By sailing west through the D'Entrecasteau Channel from Kettering, and up the west coast of Tasmania.

GPS Positions for Points of Interest

The Sulphide Lake	S 42° 31.348'; E 145° 40.57'
Lake Fidler	S 42° 3.40'; E 145° 40.56'

In Kelly Basin that night, the wind blew up a gale. I awoke with the noise of the wind in the rigging, but Ian was already up checking on our anchor. When he came back into the cockpit he said, '*Dovetail* must have come in late. She's dragged anchor and John is on deck resetting it.'

We went back to bed and I dropped off to sleep again. Our Avon dinghy flipped about an hour later and, once again, Ian went on deck. I didn't get up. There was no point in climbing out of a warm bunk. My capable husband could look after this job. Besides, it was drizzling and I couldn't see the point of making my waterproofs wet, too. I wondered if Shirley was outside helping John.

I lay there listening to Ian's movements overhead, envisaging his actions as he hauled the dinghy up. I could hear him catch hold of the bow and twist it around. Once the dinghy was turned over, he climbed into it to rescue the oars. Fortunately, the dinghy didn't have the outboard motor on the stern. That wasn't clamped to the rail in its usual position; it had been dead and buried deep in the cockpit locker since its death up the Gordon River.

Our oars seldom get lost by floating away. When our children were learning to row, we tied each oar on a long light cord and we have continued with this practice ever since. Although we sometimes tangle the cords, the oars can't float away when the dinghy flips as it did in strong winds that night.

On Friday the first of April, we waved goodbye to John and Shirley on *Dovetail*, and motored back to Strahan, the tourist village.

That Sunday was Ian's 58th birthday. Despite the blustery wind and rain showers, we struggled ashore, where I bought a fresh lobster for Ian's birthday treat. I had longed for lobster since I had seen the fisherman harvesting them outside Recherche Bay. However, this one had been refrigerated after being cooked the previous day and it had, disappointingly, lost much of its flavour. Nevertheless, we enjoyed our meal at the saloon table and I sang happy birthday to Ian. Our diet is fairly mundane because of our food intolerances, so it was good to be eating something different.

The bad weather eased the following day and on Tuesday morning, in fine but cool conditions, we motored out of Macquarie Harbour through Hell's Gates and back to sea. I was sorry to leave Macquarie

Harbour. We had both thoroughly enjoyed the wild river experience, despite not making it to the famed Franklin. I had loved the mixture of isolation, fabulous wild forest and the history of the early convicts and settlers. In retrospect, it was one of the best things we've done together.

Sailing on a north-westerly course, we continued our circumnavigation of Tasmania. A mild north to north-easterly wind made for pleasant enough sailing – despite the strong south-westerly swell out of the Southern Ocean, which occasionally knocked the wind out of the sails or, to Ian's frustration, momentarily stopped the boat. I prefer more gentle sailing conditions and I was keen to be away from the west coast with its deadly reputation in bad weather. I was slightly anxious but Ian had, of course, checked out the weather conditions as thoroughly as he could before we set off. I could always trust him to keep us and our boat as safe as possible.

About the middle of the next day, a mild southerly change came through accompanied by rain squalls. Clouds of black shearwaters flew up the coast with the wind. It was good to see some wildlife in the sky after the barrenness of the ocean the previous day. There were also several small albatross. The birds were probably following fish, but we didn't catch any on our trolling line. I didn't mind. The motion in the cockpit was too much for cleaning and filleting a fish and we still had jars of preserved meat to sustain us.

A second night at sea saw us well up the west coast of Tasmania towards the north-western tip of the island. For a while, conditions were really comfortable with the south-easterly wind on the aft quarter. But close to 41° latitude, the coast backs from a north-westerly direction, to a more northerly direction. This put our course harder into the wind and the further north we progressed, the further towards the north-east the wind veered. We reefed the sails and pulled them in tighter. Our ride became bumpier and spray began to fly across the deck. We also moved into adverse current – as much as two knots against us during the early hours of Thursday morning. I had been on watch until midnight and I awoke to hear Ian moving about.

'What's up?' I asked.

'We're not making any progress. There's a strong current keeping us almost at a standstill. I'm going to motor-sail.'

He went out into the cockpit. Thinking to help him, I got up and opened the engine cover. I called out that I'd turned on the cooling water valve and went back to bed to sleep some more. Ian is partially deaf and he didn't hear what I'd said.

I went back to sleep to the sound of the engine fighting the current, but awoke a short time later, my nostrils filled with a bitterly acrid smell. I leapt out of my bunk and opened the engine cover. The space was filled with dense black smoke, so I switched the motor off. Ian leaned into the cabin from the cockpit.

'What'd you do that for?'

'The engine's overheated and something's burning,' I said.

'What the hell… I know I turned the cooling water on,' he exclaimed.

It is usually my job to turn on the water cooling valve, so that is what I had done, not knowing that Ian had already turned it on. In my tiredness, I had turned the valve off again! Guilt swamped me.

'Uh-oh! I must have turned it off,' I said. 'I didn't know you'd already turned it on and I called to you that I'd done it. Now I've caused the plastic exhaust box to melt. My fault!' It is very easy to turn a flip-flop valve the wrong way. And I continued to make this mistake for several years.

'Well, it's done now. Open the forehatch to get rid of that exhaust smoke. Just as well we still have a replacement box.'

Ian came inside to look at the chart. 'There's a possible anchorage on the western side of Three Hummock Island. I'll set a course for there. It's shallow water for a long way out from shore, though.'

'How long until we're close enough?' I asked.

'About an hour – or longer if the adverse current strengthens.'

'Do you want to rest now?' I asked. I was in two minds. I needed more sleep, but I had caused the problem. I also wanted an excuse to be out of the cabin and the eye-stinging fumes.

'No thanks. You go back to bed. I'll need your help soon enough.'

I dozed fitfully, the sheet over my nose. By about 3.00am and with no assistance from the motor, we had moved out of the strongest part of the current and were slowly creeping in towards Coulomb Bay at Three Hummock Island. Ian set a waypoint for the position where he hoped to anchor. After an hour, he called me to watch the GPS for the distance from his waypoint. It was cold outside so I dressed ready to go on deck when I was needed. While I waited, I pulled up the floorboard next to the mast step and in the tiny bilge area there I located the spare blue plastic exhaust box.

By 4.00am, the GPS showed we had arrived at our anchoring position and Ian dropped the pick while I steered. In the gloom, I could barely see the land of Three Hummock Island nearly a mile away. We hoped the wind would remain gentle because we were unable to use the engine to dig the anchor in. There was little of the night left, but we went to bed anyway to sleep or doze until morning.

After breakfast, Ian changed the melted exhaust box for the spare one. My sweet husband never once berated me for my mistake. He knew it was an accident and accepted it as such. There was a problem and he needed to fix it, so he just got on with the job and I helped as best I could, passing him tools when he needed them.

After lunch, we moved out of Coulomb Bay and sailed round the point on the southern shore to Chimney Corner, where we thought there would be shelter from the forecast north-westerly. However, it looked better from the chart than in actuality. Instead of being sheltered by the large sand dunes, the wind gusted around them, swinging us about the anchor.

This north-western corner of Tasmania has three large islands and several smaller ones off its tip. We weren't going to negotiate those and their associated rocks in the dark even with the assistance of the GPS, so despite the boat swinging with every gust of wind, we stayed put for the rest of Friday, catching up on sleep, reading and writing up our diaries. Early on Saturday morning, we set off for Stanley on the northern coast, motoring eastwards against the tidal current.

At last, after sailing up the notorious west coast, we had turned the corner. Our south-western wilderness adventure was over, but there was still the settled northern Tasmanian coast to navigate.

6 Disaster in Bass Strait

On a calm Saturday morning in early April, Ian reported in to Mersey Radio that we were setting off from Chimney Corner, heading towards the tiny fishing port of Stanley. We were motoring against a half knot of current and the motion wasn't comfortable. There was a staccato-like short, sharp swell, which jerked the boat about. It reminded me of the monohull sailors' quip about multihulls – that while a monohull slops the tea out of your mug, the motion of a multihull jerks the milk out of your tea.

We soon ran into fog. The visibility was only a couple of hundred metres and the air felt cold and clammy on my face. As usual when there is no sunshine, my mood dropped. I wanted hot comfort food to warm me but the motion was too jerky to even think about cooking. There was a hint of dank odour in the air.

We were sailing using the Autohelm and both of us were on deck to keep a lookout for other vessels. Finally, there was enough breeze to blow away the fog and allow us to sail and the sky became bright blue with glorious sunshine. I was relieved because the quiet was blissful and I could talk to Ian without shouting over engine noise and vibration. He was becoming increasingly hard of hearing, but he wouldn't believe me when I told him he was going deaf.

I thought that when the port came into view we wouldn't have to enter with low visibility. I realised that I'd been anxious all the time we were fog-bound, anticipating difficulties.

It is bad enough entering a strange port without having fog as well. But then, as we were about a mile away, the fog descended again, though not as thickly as before, and my shoulders hunched as I squinted to make out the details ahead of us.

Stanley Harbour is, like many fishing ports, quite small and it's tucked in behind a strange hill called the 'Nut', which is the remains of a volcanic plug. It is quite distinctive, towering over the port and village of Stanley. We came in past it and entered from the east with just enough visibility to see where we were going. It seemed that the entire fishing fleet was in the harbour and the place reeked of dead fish.

The only place to tie up was against the piles on the wharf. By the time we had put a lot of fenders in place, tidied away the sails and lines and had lunch, our lines to the shore were taut – the tide was going out. Ian rearranged the mooring lines, added spring lines and allowed plenty of room for *Osprey* to take up the slack as the tide dropped further. We had been unaware that the tidal range was quite so big in this corner of Tasmania. It wasn't near the eight metres in Darwin, but nevertheless seemed about half of that, which is still a big tide. No wonder we had experienced so much current over the past two days.

Osprey *moored under the 'Nut' at Stanley*

Once we were sure *Osprey* was safe, we climbed up the ladder to the shore and set off to explore the small town. The main street was quaint in an olde-worldly way and most of the shops were built of timber – not surprising in a town that originated to service the timber industry. Whilst many businesses appeared to be standard small town service providers, there was a smattering of cafés and gift shops to cater for tourists. Craftspeople were obviously beginning to establish the east.

Ian spent time chatting to the local fishermen, while I bought a newspaper and after so long in the wilderness we were able to catch up on Australian and world news. As is frequently the case, nothing much seemed to have changed during the previous couple of months while we had been out of contact.

The tide was high at first light, so we departed Stanley very early, bound for Burnie. Conditions were gentle, as they had been the previous morning. We motor sailed against the prevailing current for several hours with the sails winched in tightly to stop the sharp northerly swell from jolting us about. Whilst I love the sense of freedom that comes when we are out of sight of land, I also like following a coastline and ticking off the landmarks as we pass.

It was one o'clock before a sea breeze came in and we were able to sail and troll a fishing line. However, the wind soon became stronger and gusty so that before long we were rolling in some of the headsail and reefing the mainsail.

We had a good sailing wind so Ian decided to bypass Burnie and carry on to Devonport. However, by late afternoon, the wind had veered and eased, leaving a secondary residual swell to roll us about and knock the wind out of the sails. Such changeable conditions are the frustrating part of coastal sailing and we were very glad to enter the Mersey River and tie up to a pontoon at the Devonport Yacht Club.

Once we had tidied up the sails and lines, we escaped ashore for a brisk walk – enough to clear our heads before dark. We were not interested in sightseeing in either Burnie or Devonport as we had done that previously in both towns. Rather, Devonport was just an overnight stop on our way to Launceston near where our good friends, Judy and Mike Handlinger, live. The Tasmanian north coast was not a place for sailing during the hours of darkness due to off-shore rocks and changeable currents.

We set off again soon after daylight, motor-sailing for the first hour or two before being able to turn off the 'iron topsail' and move along just under wind power. Coastal sailing is never quite as peaceful as ocean sailing. The wind continued to veer until it was astern and Ian needed to pole out the genoa to keep us moving. We can't leave the *Aries* to its own devices under these conditions, so we had to take turns steering by hand. It was not for very long though. Shortly after the middle of the day, we found the leads off Georgetown into the River Tamar. Instead of proceeding straight up river we needed to play the tides, so we anchored a couple of miles upstream at Devils Corner, a pleasant rural spot where we decided to have a rest day. The wait was convenient because the incoming tide was mostly in daylight by then, which gave us a better chance to navigate the eleven kilometres of the river up to Launceston.

When we entered the marina at Launceston, we were surprised to see *Byamee* tied up there. We hadn't seen Peter and Ruth since almost two years before. It was *Byamee* we'd stayed on in Darwin while we searched for a Toyota Land Cruiser camper.

We had some shopping to do in Launceston before meeting up with Judy and Mike to go out to their place and we split up so that Ian could do the technical stuff, like buying a spare exhaust box to replace the one I'd damaged, while I went off to the fruit and vegetable shop for fresh food. When I returned there was a pile of wet clothes and shoes in the cockpit and Ian had a very red line around the front of his throat. He was looking rather sorry for himself.

'What on earth happened to you? Who tried to garrotte you?' I asked.

'When I came back from shopping, Peter and I chatted for a while. When I climbed on board, my joggers slipped. I fell between the boat and the jetty and I was still holding onto *Osprey's* guardrail as I went down. My chin caught on the top wire,' Ian said.

I examined the mark on his throat. Although the skin was abraded from the wire, he didn't appear to be seriously hurt and there was no bleeding.

'Your pride's hurt more than anything, is it?' I asked. 'The problem is your worn out sneakers. They've no tread left. Maybe you ought to buy new ones here.'

'I might do that. Though I blame you for varnishing the gunwale,' he said. I ignored that comment.

'I suppose you've had a bit of a shock,' I replied. 'Have you had a cuppa? I'll put the kettle on.'

In his often understated manner, Ian wrote in his diary: 'I slipped on gunwale to mid-calf in water. Mild hanging on top guardrail wire.'

Some zinc cream on the injury worked wonders overnight, and the red line was much less noticeable in the morning.

Mike was at work at Launceston TAFE that day, so in the afternoon we walked to meet him and he drove us home. I always really enjoy catching up with Judy and Mike. They are like family and we don't have to put on any pretence.

I first met Judy in January 1970, a few days after I moved to live in Australia. I had booked into an Anglican hostel in Glebe Point Road and was assigned a shared room. Judy was my roommate. My first view of her on my second morning in the hostel was of a tousled, dark-brown curly head buried in the pillow. I went out and when I returned she was gone. This happened for several mornings, until one afternoon I was sitting up on my bed holding a glass of forbidden sherry when she burst into the room.

'You must be Judy,' I said. 'Want one?' I held up the glass of sherry.

'Don't mind if I do,' she said. I got up and poured some sherry in my spare glass.

'Cheers,' we clinked our glasses and that was the beginning of a very long and beautiful friendship. Before that Friday evening was over, we had decided to get out of the hostel and find a flat together where alcohol was not taboo. After pouring over the advertisements in the *Sydney Morning Herald* the next morning, we found ourselves a flat not far away, and within a few days we'd moved in.

As usual with such good friends, we all just pick up where we left off. We treat their home just as we would treat our own and they do the same at our place.

On Saturday, we all went back to town to visit one of the wonderful art galleries in Launceston. We were walking towards the entrance when my phone rang. It was Tilly Kanngieser calling from Scotland Island in Pittwater to tell us that her husband, Horst, had died. He was in his sixties and we knew he'd had a serious heart condition for a few years.

We were due to fly home for the wedding of Laura Anthony's younger sister. Jamie had left Laura in Panama to go home alone. I rang Virgin Airlines to see if we could alter our tickets so that we could attend Horst's funeral as well, but I couldn't make sense of what the woman was telling me. We gave up on that idea and just attended Claire's wedding. It was a shame not to attend Horst's funeral, though. Like us, Horst and

Tilly had sailed around the world together. We had first met in Durban in 1975. He and Tilly had been our good friends for many years.

We were in Sydney for a week. Jamie flew home from Cartagena (Columbia) to see Laura and to be at her sister's wedding. Jamie had balked at coming through the Panama Canal and had returned to Colombia, taking passengers from the San Blas Islands. Laura was upset that he and his boat were still in the Atlantic.

Jamie and Laura Anthony 2005

While we were in Sydney we drove to Church Point to see Tilly and her daughter, Anya, to offer our condolences for their loss. David and Jamie, too, were saddened by Horst's demise. Horst had been a frequent presence during their childhood and our families had often celebrated Christmas together. Now Tilly had lost her long-term partner and sailing companion, Anya and Paul their father.

That same day, we also drove north to see David at Lake Macquarie. He was living on his barge in the far western corner of Kilaben Bay on the Rathmines shore. We were impressed with David's progress in renovating the barge into a floating house for himself. We stayed with him for a couple of days, returning to Sydney to stay at my sister Shelley's unit on the eve of the wedding, which was to take place in a Maronite Christian church nearby at Macdonaldtown.

Laura was bridesmaid. With her dark hair and olive complexion, she looked stunning in a scarlet dress. Jamie borrowed a suit. It was very strange to see him dressed up instead of in his usual ragged T shirt and scruffy shorts. After a week of social whirl, we were both ready to be back on *Osprey*. It was already late April when we returned to Launceston.

The Mitchells at Claire Anthony's wedding in the Maronite church

I always like to visit the fishermen's supplies store in that town. I found a new waterproof coat and trousers there at a respectable price. Fishermen's waterproof clothing doesn't need to be rinsed in fresh water every time salt dries on it and the fabric is tougher – more suitable for the kind of sailing Ian and I prefer.

We stocked up the boat with supplies, fresh water and spares ready to return to the mainland. On Monday 2nd May, we left the marina and motored downstream with the outgoing tide. Five hours later, we anchored across the river from the Comalco wharf, five kilometres from the mouth of the Tamar.

For the next two days there was a light northerly blowing, making it difficult to sail north, so we rested after the hectic few days getting ready for departure. Thursday provided more favourable sailing conditions and we motored out of the Tamar and back into Bass Strait.

I always feel a mixture of sadness and joy at the beginning of a passage back home. On the one hand, it means that an adventure is almost over. On the other hand, a homecoming is also good, being a return to friends and family.

In the log, Ian reported less than two metres of swell before he went to bed. I wasn't feeling much anxiety about crossing Bass Strait. This was our sixth crossing and we were old hands at it. It was only the second crossing in *Osprey*, but she is such a seaworthy vessel with her deep, heavy keel that she inspires confidence. Also, I had been on board for long enough not to expect any more seasickness this voyage.

Ian called Tamar Sea Rescue to inform them that we were back at sea on a north-easterly course to take us to the village of Lady Barron on the southern shore of Flinders Island. We reefed and finally hove to for the passing of a south-westerly front. After sailing overnight, we arrived in Adelaide Bay late on Friday morning. We picked up the mooring that *Taipan* had just vacated, waving to them as they motored out of the bay. It was our second visit to the island. We'd cruised in Bass Strait with our children in *Realitas* during January 1986.

We lazed away the afternoon, dozing, reading and catching up on journal writing. On Saturday morning, we went ashore, and eventually managed to find a hire car for the weekend. We drove the length of Flinders Island visiting Whitebridge, the township in the middle of the island, and later driving as far as the tiny settlement of Killikrankie, where we'd previously stopped in *Realitas* and spent a happy couple of hours among the orange, lichen-covered rocks, fossicking for Killikrankie diamonds, the local topaz. The return drive down the centre of the island was on a pleasant winding road among the hills.

On Sunday morning, we drove over to Trousers Bay where the Mount Strzelecki climb starts. Although it was cloudy when we set off, the sky soon cleared. We were keen to really stretch our legs after so many days of sitting in the boat and then in the car.

At 8.00am, the morning light showed the 756m granite massif towering over the landscape. Although the track itself is not difficult, it is still a steep climb. We reached the top in two and a half hours – not bad for a 60 year-old bloke with Chronic Fatigue Syndrome and his 61 year-old wife!

At the top we saw and photographed the fabulous views over the smaller Bass Strait islands that we'd missed last time, when the clouds blocked everything a minute after we'd arrived at the summit. This time, we enjoyed our picnic lunch in perfect sunshine and with a backdrop of turquoise ocean dotted with idyllic islands. It looked the perfect sailor's playground. How deceptive such ideas can be. Even in summer, those waters can be icy.

View west from Mt Stryzlecki (Flinders Island)

On our way back to Lady Barron, we visited Wybaleena, the west coast location where early European settlers had a leper colony. A few Tasmanian Aborigines passed the final days of their lives there and for some years, Truganini was also there. In 1873, when all the others had died, she was taken back to Hobart, where she died in 1876. I am not sure of the accuracy of the claim that she was the last full-blood Tasmanian Aborigine. Her island home must have been a grim place, especially during winter when westerly gales swept in from the Southern Ocean, yet that was hard to imagine on a bright sunny day with a gentle breeze moving through the long straw-coloured grass.

The next day, the weather looked good for crossing Bass Strait. A cold front had passed through overnight and the wind remained in the south-west. We sailed out of Adelaide Bay in the middle of the day, the wind behind us. Very soon, *Osprey* would be behind the shelter of Flinders Island. The wind was expected to be squally but not strong enough to produce heavy waves and swell. We were quite comfortable about our timing for the weather and our coming passage, expecting it to be relatively quiet and uneventful.

By 10.00pm, the remaining squalls had blown themselves out and the distance between the seas and swell were lengthening. *Osprey* was sailing easily on a beam reach. Despite my confidence not to be seasick again, I was. I took watch for a couple of hours. Ian took over from me just after midnight and there was no indication of any change in the wind or sea. Our course was NE towards Cape Howe. We expected the wind to continue easing from the west and the seas to smooth out as we sailed away from the shelter of Flinders Island and into the open sea. It was with enormous shock that I next awoke in mid-air. What the …?

I wrote the following magazine article, published in *Cruising Helmsman* in January 2006.

Down but not Out

At 0220 hours, in the middle of a dark overcast night and with no warning, *Osprey* was lifted up and dropped down the face of an unusually steep wave. There was a tremendous noise and shuddering as she hit the water on her starboard side followed by the roar of water as the wave broke over us. Then we were upright again. My sleeping body had been lifted over the port lee-cloth and deposited onto the floor, along with most of the books from the shelf above and various items of food.

'Oh, shit! The boat's broken!' I cried out, as water swirled around me.

Just 12 hours before, we had departed the small harbour of Lady Barron at the southern end of Flinders Island. We'd had good assistance from the tidal current and the wind was behind us, so that by 1500 hours, we were clear of all the shallows and sand banks and had set a course for Cape Howe in far south-eastern Victoria, about 175nm distant. Weather conditions: The forecast was for westerly winds of 20-25kts, gusting to 30kts, with a two-metre swell and three-metre waves. The wind and seas were actually slightly less than forecast; we certainly would not go out in those conditions if we were heading in a westerly direction, but it was as good as we were going to get to sail north-east.

The front had passed before we left, but the wind was forecast to continue from the west because of an almost stationary high-pressure area in the Tasman to the east of Tasmania. The wind was just aft of the beam, and

Osprey A, our Brolga 33, was handling the conditions comfortably — better than I was.

I was seasick yet again and mostly keeping to my bunk. The *Aries* was happily steering and Ian was doing the bulk of the watch-keeping. Ian and I have been sailing together since 1975, and this was our sixth time in Bass Strait. We were making good progress with the sails well reefed down, doing just over six knots. There was no sign of current. We had covered about 65nm since clearing Flinders Island and were already out in the eastern, deeper portion of the east Bass Strait area, when our world was so rudely shaken.

'Where are you, Ian? Are you all right?'
A groan came from the floor close to the companionway. 'No, I'm not all right. I've hit my head and I think I've got a concussion.'
'The mast's gone,' I told him.

The foot of the mast is out of the mast step

The navigation table light was on, so we were not in pitch black. I could see the foot of the broken mast swinging about near my feet. The water had swirled up the starboard side of the cabin but, fortunately, all our electrics, batteries, radios and inverter are on the port side.

I got up from amid the debris and water and looked at Ian. There was blood on his face. I found a hand towel, which I gave him to wipe his face. With Ian injured, it was up to me to get the water pumped out and to find out whether we were sinking or not. I tried not to think about the effort of finding emergency gear and getting to the life raft with everything in such a mess.

I edged past the swinging base of the mast to the pump for the forward bilge and started pumping. On the switchboard, I could see the light go on, indicating that the electric bilge pump for the aft bilge under the motor was working. I pumped and pumped until the water was below the floorboards. Each time I stopped pumping, I started shaking with shock, so I just kept pumping until the water was gone.

Helping Ian

When I realised we were not sinking, I started picking up books, throwing important navigation books and manuals into the forward cabin where there wasn't any water. That gave me a bit of foot space to get back to Ian.

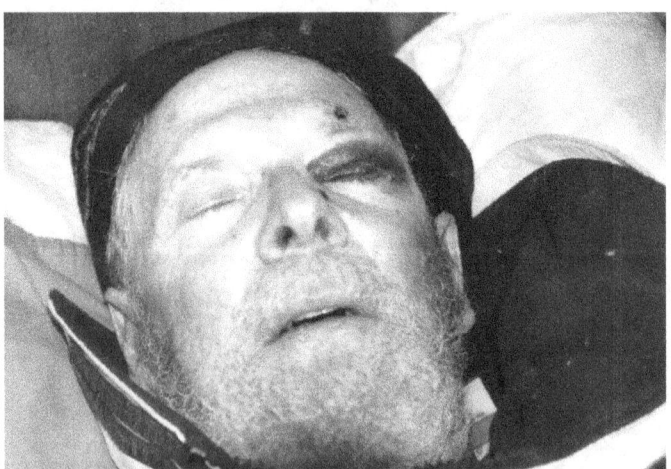

A flying fire extinguisher hit Ian, blacking his eye

The cushion was still on the port bunk, although all the boards had disappeared from under it. However, the three gel batteries are held down under there, so there were plenty of supports across the locker. I helped Ian get up and onto the bunk, tucking him in behind the lee-cloth and wrapping him in a quilt – a wet one, as everything I could get to was saturated – and a thick woollen jumper around his head to try to stop heat loss there.

He was shaking and blood was dribbling from his nose. There was a wound above his eyebrow and his left eye was starting to swell up. After a couple of hours, Ian stopped shivering, got up and went to the toilet. That relieved me greatly too. (With a concussion, there is a danger of urinary retention.) I was also wet and cold, so I crawled under another wet quilt and once I stopped shivering I slept for a short while.

The damage on deck

When daylight came, I looked outside. During the night, by torchlight, I had seen enough to know that the stern arch was still there, though mangled, and that the VHF aerial was still attached. Now, with the sea calmer, I clambered out and surveyed the damage.

The targa arch is mangled

What a mess! The mast was bent over at deck level and so was the furler. The mainsail had a huge tear in it and the genoa was wrapped around under the hull. I could see part of it on the port side, under the water. The dodger was flattened, but most of the canvas appeared to be there. We saw later that the dodger had sustained very little damage, because the stainless-steel supports had just collapsed.

I tried calling on the VHF radio, which still had an aerial on the aft arch, but I got no response. I didn't know whether the aerial was working or was damaged. There was the option of setting off the EPIRB, but we were not sinking, the sea conditions were continuing to ease and Ian seemed to be out of immediate danger.

Dismasted 11.05.2005

First job

The first job was to clean up enough to be able to move about the boat and access the tool locker. I sorted what I could, a little at a time, and managed to heat some food for Ian for breakfast. He ate a small amount, despite feeling nauseous – mostly from swallowed blood. I was still seasick and had eaten nothing since breakfast the previous morning.

I was feeling weak and tired so I lay down for long periods. I felt awful. My home had been turned almost upside down and my beloved husband of 34 years was injured, and I wasn't sure how badly. He was having difficulty understanding or remembering what I said to him. I wasn't sure what to do to clear away the rig, but thought it probably involved bolt cutters, the hacksaw and a sharp knife to cut the tangled lines.

The rigging

By early afternoon, I felt strong enough to open the tool locker and start looking for what we'd need. Ian decided that if I got the tools out, he'd try to clear the damage away himself. We knew that we needed to get under way before dark, but we didn't know if the engine had moved on its mounts. It was too dangerous to try starting the motor until we had made sure there was nothing near the propeller.

It appeared that the mast had fractured under the spreaders from the compression force of the water, then bent over at deck level. At times like this, I am thankful for Ian's ability to stay single-mindedly focused to the exclusion of all personal comfort. He stayed on deck for two-and-a-half hours until he had released the last wire, sending the whole mangled mess of rigging sliding down to the ocean depths. The boom had been torn off at the gooseneck fitting and was also bent. The furler was so twisted it seemed a waste of effort to try to retrieve it.

We made sure there were no loose lines overboard and then prepared to start the motor. Yes, it started! We left it running for about ten minutes before Ian dared to put it into gear. There was no irregular thumping. Nor was water pouring in through the prop shaft seal. Hallelujah! The propeller was not damaged, nor the prop shaft. As soon as we were underway the motion smoothed out. Exhausted, Ian went to bed for a few hours and then relieved me at the tiller so I could get some sleep too.

Contact

By mid-morning the next day we were in sight of Cape Howe and – although we could still raise no-one on the VHF – we did get a weak CDMA signal on Ian's old mobile phone. I called our younger son, David, who was very concerned no-one else knew where we were. He phoned Eden Coastal Patrol to let them know our plight. They, in turn, rang us. They promised to call Mallacoota Coast Guard in south-eastern Victoria and alert them that we were entering their area.

I felt enormous relief that our position was now known, and someone would be looking out for us. The day seemed brighter.

We motored on. Ian had drained the dregs from the bottom of the fuel tank before leaving Tasmania, so the fuel filter didn't clog up, a distinct possibility after such a severe shaking. As we approached Gabo Island, we radioed Mallacoota Coastguard and they answered. Now we knew that the VHF was working and I wondered why they hadn't answered before.

Twofold Bay

Two hours after dark, we entered Twofold Bay. Ian was still feeling ill and we decided to anchor behind the woodchip mill, rather than try to wend our way through moored obstacles to anchor off the fish cannery. We followed the chart and the GPS to get in behind the wharf to the small bay where yachts often anchor.

Despite having been in Eden many times, we had been into this hideaway anchorage only twice before. Now we edged into the bay and Ian went forward to run the anchor out while I managed the engine. The anchor went over the bow and dropped to the bottom, and then it would go no further. The chain was tangled and no amount of tugging would free it.

We were so tired. I emptied the lazarette to find the danforth anchor. Then I dug out the nylon anchor warp while Ian emptied the tool locker. The spare anchor chain lives in a bucket next to the hull, in the level under the tools. Finally, I hunted for shackles and spanners while Ian hauled everything on deck. An hour later, the spare anchor was finally holding *Osprey*, and we retired inside to eat and sleep in our still-wet bunks.

After breakfast on Friday morning, we motored across Twofold Bay and tied up to a fishing boat at the wharf. I went ashore and phoned Club Marine, with whom we are insured, to notify them that we would be making an insurance claim. Just explaining what had happened caused me to become teary and start shaking again with shock.

In town, I came across a physical practitioner who helped enormously with the shock to my body. I booked Ian in to see him that afternoon. After that, Ian decided not to go to a doctor because his headache was decreasing and he had no other symptoms of head injury.

After several days recovering in Eden and enjoying the hospitality of the wonderful men and women of the Royal Volunteer Coastal Patrol and other cruisers, we motored on to Sydney for repairs.

We did it!

We had a degree of exhilaration as we headed home. Despite the damage to ourselves and the rigging, we'd saved our boat and brought her into port ourselves. What was more sobering to me was the knowledge that even a relatively small yacht was still too much for me to handle on my own in an emergency, although it's amazing what you can do when there is no option and your adrenalin is high.

What we learned from the experience

Secure everything whenever you go to sea. We had left the rail for the bookshelf, the fire extinguisher and the companionway ladder unsecured and a gimballed oil lamp in its holder. The dinette table had inadequate hold-down bolts. Further, the hold-downs for the tool and food lockers were also inadequate. To our shame, we hadn't even bothered to do them up.

As a result of these inadequacies, the books all came off the shelf, lighter food items jumped out of the food locker and the companionway steps turned themselves upside down and back to front, piercing the cabin ceiling.

The oil lamp became a missile, holing the cabin ceiling. The fire extinguisher was the object that hit Ian. It could have killed him. The table top dislodged itself in such a way that access to lockers next to the table was impossible. I hate to think what would have happened had Ian actually been outside. He was on his way out to look around, and was about to step onto the ladder. Thirty seconds later, he would have been standing in the cockpit. Would he have bothered to snap his harness strop onto the U-bolt in the cockpit for those few moments outside?

Both of us had become careless about such matters. I feel sure he would have been much more severely injured or lost at sea had he been outside. We definitely received a safety wake-up call.

Ian added the technical addendum:

Rogue Waves

The *Maxwave Project* defines a rogue wave as 'an individual wave of exceptional height or abnormal shape'. As part of the *Maxwave Project*, radar images were taken of 30,000 areas of ocean, each 10 x 5 kilometres. Ten waves of more than 25m were observed. Waves of abnormal height are identifiable on radar, but there is probably no current technology to identify waves of abnormal shape.

The usually accepted criterion is that, no matter how stable the yacht, it will be rolled by a steep breaking wave on the beam if the wave exceeds 55 percent of the length of the yacht. Thus an abnormally steep and breaking wave of five metres would be sufficient to drop *Osprey* down its face, then break over us. In my opinion, the chance of being exactly abeam where such an abnormal wave is starting to break is extremely rare — perhaps similar to the chance of being hit by lightning out at sea. (Ian Mitchell)

⚓

The insurance company advised us to take *Osprey* back to Sydney. Later, we wished we'd gone to Melbourne. However, because the weather was calm, and Sydney was where friends and family lived, that's where we went.

We assumed the insurance company would be difficult to deal with, but once we'd had an insurance assessment completed and had received the go-ahead for repairs, we believed that it would be a fairly quick job to have *Osprey* fixed. Instead, the insurance company personnel were sympathetic and helpful, but the rigging company, in particular, was difficult and the pipe bender for the arch even more so. Dealing with these tradesmen was almost as stressful as the rolldown itself.

At some stage, we recognized that the screws for the handholds running along the deck-head had been bent. There was also the fact that the side of my face looked like it had been swiped by a cricket bat. I found, when I looked in a mirror in Eden, that my left jaw was black and blue, later becoming purple and yellow. In retrospect, it was fairly obvious that the side of my face had impacted the handhold before I fell back to the floor while one of my legs was still caught up in the lee cloth. Far from being uninjured as I had thought, I too had a head injury and part of my hamstring was torn off the back of my pelvis.

Eventually, the whack across my jaw led to the loss of a back molar (the root died), I found that I no longer had filtering for high pitched sounds in my left ear and I had suffered some brain damage. None of my injuries was easily observed by others once the facial bruising had gone.

It became obvious that not just *Osprey* needed fixing. Both of us did too, and that took many months.

7 The Aftermath

Our first job, after establishing ourselves in Sydney harbour, was to visit the insurance office. I'd had my film printed and took along an envelope full of photographs of the damage to *Osprey*. The insurance people were highly sympathetic about our accident and assured us all would be processed relatively quickly, so that we could get on with repairs.

In yachting circles one often hears stories of people having difficulties with their insurers and of rather unhelpful assessors, so we were sceptical. However, only a few days later, the insurance assessor contacted us to arrange for him to visit us on board. In this instance, the gentleman could not have been more helpful. After the trauma we'd experienced, and that I was reliving every time we had to go through the details, we were keen to have our boat restored to a seaworthy condition as soon as possible. We needed to go out to sea again, to reassure ourselves that it was not our fault, that the wave was a rare occurrence and that it was safe to go cruising again. The assessor seemed to understand our point of view that a tough vessel like the Brolga was worth repairing.

We discussed which aspects of the repairs the insurance should cover and which we could do ourselves, so that the total came out at a figure just under that at which the company would declare *Osprey* a write-off.

We set about making enquiries from ship yards and repairers who might give quotes, and were astonished to find that these businesses did not return our calls. It was winter, after all. How could they be so busy at that time of the year? Having started our enquiries enthusiastically thinking we would have a choice, we were reduced to selecting those who had space in their schedules.

We knew Five Dock Bay well and it was not too far to drive to my sister's place. We knew where to park our car so that it was safe from vandals and so we had anchored not far from where we'd anchored *Libelle* four years before. But someone reported us again and it wasn't long before the Waterways boat visited, telling us we had to move into a marina. We hired a mooring in Drummoyne instead and moved into my sister's small unit.

'It will be for only a few weeks, Shelley,' I told her.

'Okay, if you do the cooking and washing up,' she said.

'It's a deal.'

Our living with Shelley was not a good situation for her. She likes her privacy and her home is a small, one-bedroom unit. We slept in her living room on a futon lounge that was less than ideal for my injured lower back.

On the positive side, we could use her washing machine and we were close to the train line if we needed to go into the city. There was also a tenant in the building without a car, so we were able to use the vacant off-street car parking space.

Eventually, we found a good shipwright at Balmain. His business was close to the mooring at Drummoyne. We moved *Osprey* onto one of his moorings and employed him to fix the damage to the fibreglass cabin top, to repair the damaged woodwork and replace the broken fibreglass stanchion bases with stainless steel ones. He was an excellent craftsman and we were well pleased with his work.

Finding a rigger was the real problem. None of our preferred firms would even quote, and it was late July before we found one who was willing to do the job for us. Ian had drawn up revised rigging plans.

Joubert's designs always included a steel ring frame which interconnects the chain plates and the mast support, and is an integral part of the structure of the vessel. Hidden by cupboards, the frame passes down the sides of the cabin top, down the inside of the hull, under the floor and the foot of the mast, then up the other side of the boat. The chain plates hold the shrouds which support the spreaders.

Ian wanted a double-spreader rig and a modification to the chain plates on each side of the cabin. The extra set of spreaders needed two more holes for its shrouds. The chain plates are bolted to the ring frame, completing the integrity of the rig and hull. The old plates had to be removed and the new ones that Ian had designed and ordered from a fabricator had to be bolted on with sixteen new bolts. In the process of removing the original chain plates, Ian found that all of the nuts had been epoxied onto the old bolts and six of them broke when he tried to remove them. He had to drill out all of those bolts.

Ian and I had endlessly discussed the new sail plan. We were absolutely positive we wanted to have three point slab-reefing for the mainsail. We also wanted a genoa on a furler that didn't hug the deck, so that when we were sailing in crowded harbours we could see under the sail. Deck-hugging genoas might be good on a racing boat with six crew members but it didn't suit our style of sailing. I preferred the high cut of a yankee.

All the stress gave Ian a resurgence of Chronic Fatigue Syndrome, yet he insisted that I should not deal with these tradespeople.

'These guys are used to dealing with men,' he told me. 'You don't have the technical knowledge required.' He also thought my manner too abrasive. Probably true.

Ian refused to listen to my protests. It was unlike him and I was frustrated and indignant. I wanted to help explain what we wanted and that Ian's CFS required him to sit down, to argue for his viewpoint when he became brain-fogged, but I was shut out and kept in the background. It was half my boat, damn it. But for once, Ian insisted. Consequently, as his brain-fog increased, I feared he would agree to things he might not have in normal circumstances.

When the rigger came on board initially and sat at our little dinette to discuss Ian's plans, I tried to suggest he move over and let Ian sit down too, but I was ignored by both men. When we visited the factory, I wanted to ask for a chair for Ian because he couldn't remain standing for long. Again, I was shushed.

However, Ian did reject the rigger's idea that we needed to run only two reefing lines for decreasing the size of the mainsail.

'How would that work?' asked Ian.

'You pull down the first and second reefs in the normal way. You untie the first reef, thread it through the third cringle and, using a light line, pull the reefing line through the third reefing points, tying off on the boom.' His assumption was that we'd never need it.

Ian says he stared at him, envisioning the perilous manoeuvres required. Surely he must be joking, but the rigger remained serious.

'No. No, that is definitely not on,' said Ian. 'First, I am used to reefing the main at any point of sailing. I don't want to have to pull the boom in hard just to reef. Second, I don't want the third reefing line coming from the aft end of the boom. It is impossible to get enough tension to pull the third reef cringle down to the boom. The boom will sag down. I want the third reefing line coming out of the side of the boom to a cheek block on a track so it can be adjusted to the correct angle. 'Third, standing up on the cockpit seat and using both hands to fiddle with lines is just ridiculously dangerous. No way. My wife would mutiny if I tried that.'

Little did he know his wife was already mutinous. A few months later, Ian showed me the cover of *Australian Sailing* magazine. The photo showed the cockpit of a 50' racing boat. One of the crew was standing on the cockpit seat with another crew member on his shoulders so he could reach to put in a third reef, just as the rigger had suggested. We were both aghast at the risk this crew was taking.

The rigger did as Ian asked, but the track and third reef block on the side of the boom were not at the correct angle, so that the block chewed the line. Later, Ian replaced both so that they worked properly.

The required mast section was available only from Victoria and the factory there had acquired new extrusion machinery. To extrude our mast shape, the die had to be modified and that caused the first delay. When the mast was finally extruded and arrived in Sydney, the rigger had to measure and cut all of the holes for wires and fit all of the attachments, such as cleats, goose-neck for the boom, ring for the spinnaker pole and mast-head for lights and aerial.

Once all the screw holes were drilled and all that fitting was done, everything had to come off again for anodising the mast. There was no anodising bath large enough in NSW, so the mast had to be sent back by truck to Victoria to be anodised and eventually returned to Sydney.

We were not preparing for the racing season, so by August any racing boats took precedence over us. It didn't matter that we wanted to be back living on our boat and had plans to cruise to New Zealand that summer. There was one delay after another. In early December, the mast had finally returned from Melbourne and everything was ready at last. The rigger booked a berth at a place in Manly for the new mast to be stepped.

Ian and I were excited. At long last we'd have our boat back, just in time for Christmas.

'I want to go up to Lake Macquarie to have Christmas with David,' I said. 'I think we might manage that,' agreed Ian.

We motored *Osprey* to the wharf, which had a small fixed crane for lifting masts, and tied up alongside. Soon after 9.00am, the rigger's truck arrived and the men offloaded our new mast. Ian helped them carry it down the path from the road to the wharf where they set it down. Then they brought down the new furler and lay it alongside the mast.

I was full of expectation. The gear was here at last and the riggers were finally going to install our mast and tighten up the rigging. *Osprey* was going to look like a sailing vessel again.

'After today, we'll be able to go sailing,' I said to Ian.

'Don't forget we have to have new sails first.'

'Oh, how could I have forgotten?' I felt deflated. I really had forgotten about the sails and Ian hadn't meant we could *sail* to Lake Macquarie.

Ian removed the jury rig and prepared the deck for the new mast. It seemed the men took ages to set the mast up ready for the crane to lift it and position it down through the new mast collar at deck-level and into the mast step on the ring frame under the floor.

All seemed to be going well. The fore and aft stays had been installed and tightened. After lunch the team unwrapped the shrouds and stays that were attached and lashed to the mast. The shrouds were a whole metre too short to reach the chain plates. The rigger had made a simple error in his measurements.

I was fuming. I felt frustrated and helpless. It was all so incompetent and wasteful of time and money. I wanted to tell them what I thought of them, but Ian wouldn't let me. There went my plans for taking the boat to the lake for Christmas!

The riggers had to lift the mast out of the boat and store it along with our furler until they could make another booking for the crane. Sadly, we motored *Osprey* back up the harbour to her hired mooring in Drummoyne and went home to Shelley's.

'That rigger had better get it right next time,' I growled as we retraced our passage to Manly and the wharf with the hoist a couple of days later. The riggers returned to the wharf with the mast and furler and prepared the crane to hoist the mast into the boat.

'I hope the shrouds are the correct length today,' I said to the rigger. I couldn't stop myself and Ian was out of earshot.

'I hope so too,' he replied.

The riggers raise the mast

They were, and this time everything fitted but for one detail. All the wires for lights and other electricals were fitted inside the mast. Ian had asked for them to exit at a point just below the cabin ceiling. Instead, they exited down near the mast step. There was another delay while that was fixed. At long last, everything was done: the halyards led inside the mast, the shrouds were done up and the furler was attached.

The day was done by the time the crew had finished everything. Ian and I were weary too. It had been a long day. Ian wrote in his diary:

'Mast in!!!' It was the 9th of December 2005, seven months since the accident in Bass Strait. Our struggles to find competent tradespeople seemed to have taken forever.

During the previous couple of months, Ian had been fighting another battle. He drew up plans for a new arch for *Osprey's* stern to hold the solar panels. The tradesperson Ian hired didn't get back to us for a few weeks after accepting the job. When Ian finally managed to contact him he said,

'My pipe bender has just retired. I'm looking for someone else.'

The next communication from him amazed Ian. 'My new fellow says he can't make the bends you've asked for. He says he can't make two bends in a pipe.'

'I'll talk to him,' said Ian. He turned to me and said, 'What kind of pipe bender can't resolve angles in two planes into one angle?'

Later, on the telephone, Ian asked for details of the pipe-bending machine. He talked the fellow through the process of bending the pipe to the angles he needed. Later, the man rang back. 'I've done the left side, but I don't know how to do the right side.'

After talking him through the reverse process, Ian put the phone down. He was laughing.

'I just can't believe this fellow. What sort of training has he had? After I've told him how to do one side, he's incapable of transferring the instructions for the opposite side!'

We finally got our targa arch onto the boat and installed it, relieved that it did fit. We even saved our original two solar panels and bolted those on again, although the older one needed a lot of *Sikaflex* to make it waterproof.

The dodger required only a very small amount of work. It has a fold-down framework, which had collapsed under the weight of the water. I removed the vinyl cover from the framework and examined it.

'Jos Bots might be able to fix it for us. I think he has an industrial sewing machine.'

'Yes, Jos does. He makes his own sails and I'm sure he makes awnings and dodgers, too,' said Ian. Jos belonged to our cruising club and was therefore easy to contact. He was very obliging and after I dropped the dodger off at his home, he repaired it quickly and efficiently.

The new targa arch on Osprey's *stern*

That made two repair jobs performed with little fuss! It was just the major repair of the mast and rigging that brought huge frustration.

During January, we were finally able to call in the sailmaker to measure for replacement sails to fit the new rig. Ian Short came on board and made his measurements. Once again, Ian ordered me to stay in the background. While I grumbled about this, Ian agreed to a low cut genoa for the furler and full length battens for the mainsail. The genoa did stay above the pulpit rail, but it wasn't the high-cut yankee Ian and I had talked about. I was unaware of these decisions until Shorty came back with the finished sails. The two Ians fitted the new sails onto the forestay and the boom and we went for a short sail.

Ian hadn't explained to me that he had changed his mind about a yankee, so I was horrified. The sails were made of heavier cloth but, even so, it seemed to me that we now had a pair of beautifully made racing sails. When the genoa was fully unfurled, there was no view ahead of the boat at all unless one of us went forward and peered under the foot of the sail – something I had wanted to avoid.

The slides for the full-length battens on the mainsail provided resistance on the mast track, which was annoying when we were raising it. If I didn't pay absolute attention to *Osprey's* angle to the wind, it could catch those battens and flip them back into the opposite curve. The whole sail was very heavy for me to pull up the mast, even using the winch. I preferred the style of mainsail we'd had on *Caprice* where the luff was cut on a reverse curve so it didn't need any battens at all, but battens allowed for the sail to be larger in area. Having a full cut genoa instead of a yankee with a high clew also provided greater sail area. As half owner and sailing partner on the boat, I was very indignant that Ian hadn't discussed these matters with me. More inexplicable was why so much effort should be needed to hoist the dinghy from the water onto the deck with the spinnaker halyard.

Osprey *is finally repaired and has her new double-spreader rig*

Ian climbed the mast and found that two of the sheaves at the top of the mast weren't turning – the one for the spinnaker halyard and the one for the genoa. The sheaves were jammed onto the top of the masthead section and the ropes were cutting grooves into them. No wonder it was so difficult to pull up those halyards. He took photos of the situation and sent them to the rigger, who posted us two new sheaves. Ian ground them down by 4mm and then installed them. They were then free to turn properly.

The final straw came when we tried out the slab-reefing. We discovered that the line for the third reef had been cut so short it wouldn't even reach the cleat on the deck, and Ian had to splice an extra half metre of line to it.

It was early February before we could move back on board and sail our boat up to Lake Macquarie where we found there was a very competent rigger and sailmaker!

Repairs to our bodies were also on the agenda at that time. I went to an acupuncturist. The elderly Chinese gentleman asked me when I had broken my nose.

'As far as I know, I've never broken my nose,' I indignantly replied.

'Well, your nose has been broken at some time,' he insisted. 'It is difficult for me to help you become well when you are not able to breathe properly.'

I ignored the wise old man, only to remember some years later that I had cracked my nose against a bulkhead on *Caprice* in 1977. Instead of just sustaining severe bruising as I had thought at the time, I must have broken the bone.

My pelvis was causing me pain and grief too. An osteopath's ministrations left the condition worse. Ian had recurring problems with his back. We consulted an excellent physiotherapist in Artarmon.

'I've seen this condition before in long-distance flight attendants. When the plane drops into an air pocket, a sleeping hostess can find herself thrown out of her bunk. If her foot is left behind in the bedclothes, as yours was in your Bass Strait mishap, part of the hamstring gets torn from the pelvic bone.'

'So that's why I feel as though my pelvic organs are unsupported!'

'Exactly,' she replied.

After treatments, my hamstrings started healing themselves and pelvic floor exercises were effective once more.

Ian's treatment was also effective. We were both repairing ourselves ready to go cruising again.

As for our next cruise – we had missed the opportunity to sail to New Zealand during the summer of 2005-6. We would have to tackle that another time. We decided instead to prepare for a tropical voyage to New Caledonia and Vanuatu during the coming winter.

8 Voyage to New Caledonia

New Caledonia has always seemed a romantic place to me. I first heard of the islands in 1963 when I was in senior high school studying French. Nouméa, the capital, was the nearest place to go from New Zealand for students to practice oral French. I had longed to go there too.

Now Ian and I began planning in earnest to sail there during the winter of 2006. We hadn't sailed to a foreign port for many years and the thought of doing so was very exciting. We decided to leave from Coffs Harbour. The direct distance from there to Nouméa is under 1,000nm, but of course, when sailing, one covers a much greater distance. We estimated it might take us about ten days to arrive.

There was an enormous amount of work to do before we could contemplate setting off. Before we make any passage at sea, Ian makes a mental list of all the things that could go wrong with the boat and then writes lists of the necessary spares. He decided to order a new high frequency (long-distance) radio, which had to be installed and tuned. The new sails and rigging had to be adjusted; sheet ropes that had survived the dismasting and other aspects of the running rigging needed to be measured and replaced; rope-ends had to be sealed or whipped; swivels and shackles replaced. We installed more tie-downs and handholds, new batteries and hold-downs for them. In other words, our rolldown in Bass Strait had made us even more safety-conscious.

There was an old wind generator tucked away on board. We had seen it when we first inspected *Osprey* and ignored it, but now Ian pulled it out and decided to overhaul it. We realised it was too heavy to be hung in the rigging as it was designed to be – perhaps on a heavier, larger boat. Instead, Ian installed it on a pole at the stern on the starboard side. He calculated how high it needed to be not to interfere with the flow of wind over the vane for the self-steering system, yet we still had to be able to reach it to lash it down in high winds. There was no switch to stop those blades from rotating furiously and we feared a blade shearing off and becoming a dangerous missile.

Three months passed. It was already June and we despaired of ever getting away. Ian was feeling quite exhausted, but I was still enthusiastic.

'You'll feel much better once we're out at sea,' I reminded him. I collected the meat I'd ordered and stacked it in the freezer; I wrapped fruit and vegies in paper and stowed them low, next to the hull. We said goodbye to David and the few friends we'd made in Rathmines.

That afternoon, our first stop was just round the corner at the Rathmines F jetty, where we filled up with fresh water and binned our garbage. Across the lake, we picked up the courtesy mooring inside the Swansea Bridge, but I accidentally dropped the boat hook overboard. Rescue missions are always good practice for man overboard and we managed to retrieve the hook. I realised that if we were trying to retrieve a person or even just a piece of dropped equipment at sea, it would be much more difficult to find among the swells, let alone retrieve a heavy person.

At five o'clock, the bridge opened for us. We were tired and wanted a gentle start to this voyage so, outside the bridge, we picked up another mooring and finished the last-minute stowage, making sure everything was tied down. After the Bass Strait incident, we didn't want things flying about. By the time we were ready for sea, we were both too exhausted to feel elated, but sleep re-energised us.

I prepared breakfast for Ian and we crossed the Swansea Bar at slack tide just before dawn but, as usual before going to sea, I didn't eat. I became seasick quite quickly in the squally southerly weather. Even though I turned my electronic relief bracelet up to full strength, it didn't work very well for me this time and I continued vomiting and feeling miserable, though despite the nausea, I did manage to take watch and give Ian some rest.

Osprey reached Port Stephens early that afternoon and we moored to a public buoy at Nelson Bay. Our plan was to stop there and take it easy for two days but such plans don't always work!

The toilet pump had a build-up of calcium. Toilet maintenance is never an odour-free task. Regular soaking in vinegar usually keeps things working trouble-free, but obviously we'd neglected it while we were otherwise busy.

Tony, on the cruising yacht moored next to us, was chatty.

'Have you got electronic charts?' he asked. 'No? I can lend you my CMap CDs if you have a way of copying them.'

We accepted gratefully. This was our first foray into electronic navigation. Excitement and frustration ensued with the usual computer hiccups. We copied the maps, but we couldn't open the executable file. By then, Tony had gone.

'Back to square one,' said Ian. 'We'll rely on our paper charts as originally planned.'

⚓

We weren't sufficiently rested, but with a southerly forecast the next day and the wind moving through the west, we started north again. For much of the day, the wind was off the land, making the sea smoother and sailing conditions much more pleasant during the day, but chilly overnight. I was already feeling happier to be sailing.

It was awe-inspiring to see a whale breaching the first day of that two-day sail to Coffs Harbour, and another next morning, reminding us that, in June, the majestic humpbacks migrate north. The second whale was much closer – close enough to smell its musty breath. Although delighted by the whale's presence, I also worried that it might damage our boat. Watching whales spout, jump and tail-slap is one of the joys of being out at sea, but if one surfaces too close, the tail-slap could hole the boat.

Many years ago, we met the owner of a steel boat that had suffered hull damage from a whale. Fortunately for him, there was a cargo ship with a winch close by, which lifted his boat out of the water. Another yacht was sunk by whales in the eastern Pacific in 1971. In *Survive the Savage Sea*, Dougal Robinson told of his family's thirty-eight days adrift in their dinghy and life raft after their boat was holed by an orca.

Just before dusk, the wind died altogether. We motored the last few miles into Coffs Harbour and settled into the marina. There were a couple of other yachts in the marina that were also sailing to New Caledonia. We were all eager to set off for our destination. After we had the checks on all our papers and passports completed, we had four days in which to depart. This system allowed yachties to wait for the right wind. Dealing with the customs officer at Coffs Harbour is much more pleasant and friendly than in commercial ports like Sydney and Newcastle.

Before leaving port, the last thing to do was check weather forecasts at the nearby internet café. In 2008, marinas didn't yet supply Wi-Fi. Finally, we set out to sea late on the afternoon of the last day in June – a Friday.

Sailors of old had many superstitions and many people are still superstitious about leaving port on a Friday. One theory about the origin of this belief is thought to relate to Jesus being crucified on a Friday. We've often left port on a Friday and it has never made any difference to us.

We pulled up the mainsail in the calm conditions of the outer harbour, which was fortunate because darkness had descended and the wind and sea were boisterous by the time we were outside the entrance. We were soon sailing in the right direction with the wind behind us but the bouncing motion made me very seasick again and the electronic band had no hope of warding off my nausea. I turned the pulse to its highest level, only for the nerve in my left wrist to become painful, so I moved it onto my right wrist.

All night and most of the next morning, we made excellent progress at over six-and-a-half knots. In the first twenty-four hours, despite one to two knots of current against us, we covered 130nm.

The second night, the wind dropped off and after a couple of hours of motor-sailing, we furled the genoa and pulled down the mainsail. With a calmer sea and not another vessel in sight, I was soon over the nausea and able to keep food down. I was able to really enjoy the sailing from then.

A pleasant Sunday morning with no wind tempted Ian to stop the boat. He jumped over the side to swim twice around our hull, but I wasn't so keen. With the water at about 20°, I chose to have a strip wash on board. How good to feel clean and smell fresh again after having been seasick! The wind came back in fits and starts – we rolled the sails out and furled them in again. Frustrating!

During this calm period, Ian also tried unsuccessfully to use the new H.F. radio for a telephone call to a friend in Sydney. This friend was our on-shore contact whose details we'd given to the Australian Maritime Safety Authority in the event we might use the emergency beacon (EPIRB) and call for rescue.

In the meantime, Ian had been reporting our position on Sheila Net, which covers the NE Coast of Australia, New Guinea, Louisiade Archipelago, Solomon Islands, Vanuatu and Nouméa (Sheila Net deals with emergencies, position reports, general information and traffic contacts). There was other traffic about. We heard two other yachts on a similar course to ours reporting in on Sheila Net. They were probably the two Australian boats in Coffs who had also been preparing to sail to New Caledonia.

The man who had sold us the HF radio also operated a calling schedule from Lake Macquarie for sailors and caravanners. We were unable to raise him at first but did so after a few days.

There were rain squalls – lines of dark clouds which occasionally lit up with lightning – and because the wind refused to stay in one direction for long, my nausea returned that night. This was more seasickness than I was accustomed to, and a few years later I learnt this was most likely a result of the concussion I received in Bass Strait the previous year. Apparently, brain injury can be permanent when the concussed person is over fifty-five.

Ian and I were sitting side by side on the settee when the boat heeled and then heeled further. Our eyes grew wide and we clutched at each other for a second or two. When *Osprey* came back upright, we laughed with relief. In that moment, we had both feared a repetition of the rollover in Bass Strait, despite knowing that what we'd experienced was a rare occurrence. When the wind finally steadied from behind us, we bowled along at 6-8 knots, once again making excellent progress under a blue sky and fluffy clouds. This time, my nausea didn't return.

At night, we gloried in the sparkling expanse of the heavens. To see the full wonder of the starry night, you need darkness on earth. Only in the outback and on the ocean far from shore are the conditions right. The Milky Way and the planets can be so bright that they are reflected in the water, especially if there is little or no moon. We could see more pinpricks of light than most city dwellers could ever dream of. The night sky is an added bonus to ocean sailing.

Bioluminescence is the other wonder of night sailing. Caused by diatoms in the water, when they are present the wake astern glows, glitters and gleams. Dolphins occasionally come to frolic in the bow wave and their outlines appear like fiery torpedoes creating a dazzling display as they race each other. The water hisses soothingly, swishing past the bow as the ocean sings its allure.

The winds continued to be changeable in direction and variable in strength. During our fifth night it blew 25-30 knots, not easing much before midday. Later that day, when Ian went outside to check all was well, he noticed two cargo ships passing in opposite directions across our path. We had let our vigilance slip because we estimated we were outside of shipping lanes. So much for our decision to be more careful!

Birds are yet another joy, though you see very few of them right out in the ocean. We had covered approximately two thirds of the distance to New Caledonia when we spotted a very large albatross at dawn. We were surprised to see such a bird at this latitude north, but surmised it must have been blown there in strong winds the previous day. Most of the birds we see are closer to shore and they are often the first indicator that land is near.

Our younger son, David, had given us an old multi-purpose spinnaker (MPS) from a previous boat of his and when the wind fell very light, we put it up.

We set the worn MPS David had given us.

With the Autohelm steering, we made 1-1.5 knots overnight and most of the following day. All seemed well when we sat down to eat dinner early in the evening except that the sky had developed a light cloud cover.

A sudden wind change arrived, the spinnaker filled and – with a very loud bang – blew out of its outside seams. I jumped, momentarily thinking we were being fired at. It was time to take that big light sail down, because the wind change stayed and brought several showers with it. On a positive note, we thought perhaps we'd finally met the winter south-easterly trade winds.

We had been unable to gain enough easting over the past few days and now, frustratingly, current and wind combined to push us slightly too far north. We had less than ninety miles to go to the Pass du Nord into the New Caledonian lagoon. Ian pulled the genoa in tightly and we roared along at over six knots all night.

Whilst the full moon reflected off the water and the sky was bright with stars, it wasn't a night for stargazing. There was too much water flying. When I was on watch, I would stand in the companionway, take a quick scan ahead and then duck before the spray could hit me in the face.

On the final (9th) day of our passage, just after dawn, we resorted to tacking for nearly half an hour to gain the right angle to reach our waypoint and then enter the Pass du Nord, which seemed very narrow after we'd been at sea for nine days. Once inside the lagoon, the character of the water changed as the swell dropped and the surface became choppier. Now we could smell the unmistakable odours of land – seaweed, vehicle emissions, even cooking. Our noses were ultra-sensitive after more than a week at sea.

We had almost arrived and we were both grinning with excitement. This was our first foreign landfall for many years. We gazed across at nearby L'îlot Amédée with its soaring 19m lighthouse, the tallest in New Caledonia.

'We'll come back and visit that,' said Ian. We both enjoyed climbing to the top of lighthouses and this one had its own tiny island low to the water. However, going ashore right then was illegal because first we had to clear customs, immigration and quarantine checks.

Three hours later, after a pleasant sail across the lagoon, we wended our way into Port Moselle and Nouméa, the capital and only city in New Caledonia.

Approaching the harbour entrance of Nouméa

In accordance with regulations, Ian had called the Port Captain the day before to announce our approach and now I made the calls to Nouméa Radio to declare our arrival and to Port Moselle Marina to request a berth.

Don't imagine me lolling on deck gazing at the sights. I was busy down below tidying the cabin, cooking up a pot of vegetables for our next few meals and also preparing something for lunch. We knew that all uncooked fruit and vegetables would be confiscated by the quarantine officer.

As we came closer, Ian was uncharacteristically uptight about entering the marina berth. Perhaps he was remembering the fiasco with *Libelle* at Marmong Marina when we were showing her to prospective buyers. He snapped at me a few times as I brought fenders, mooring ropes and the boathook out onto the deck, but it was all easy in the end. The marina manager came to help us tie up at the visitors' berth and a couple of Americans also gave a hand.

Ian had to visit the Port Captain's office immediately, so while he was away, I continued with my clean-up and also used the last of the eggs to make some 'bread' in the frying pan.

After lunch, we filled in the forms Ian had brought back. Customs officials came first and then immigration. By 4.00pm, the quarantine people hadn't turned up, so I went ashore to ask at the Marina office. Eventually, a very pleasant Asian lady arrived. As we'd anticipated, she took all of our remaining fresh vegetables, fruit and meat.

Entering port in your own boat is a lot more time-consuming than arriving in an aeroplane. It takes at least half a day and by the time you fill in all the forms and submit to agricultural and customs inspections and you've also done the paperwork for the marina, your excitement at arriving in a new country has evaporated. You haven't talked to so many people for days and by the time you are free to go where you want, you feel exhausted.

Our voyage from Coffs Harbour to Nouméa had taken just short of nine days to travel 970 nm. Ian calculated that we had averaged a speed of 4.6 knots, we were becalmed for 2.8 hours and we motored 7.4 hours. With the exception of when we had just left Coffs Harbour, we had experienced very little in the way of current to either help or hinder us.

After unpacking the inflatable dinghy and flexible solar panels from the vee berth so we could use our double berth that night, we had showers, dinner, and then we set off on foot, keen to explore the nearby area of this foreign city.

View of harbour from above the cathedral, Nouméa

Meridian of Sydney, with owners Paul and Judy (CCCA members), was in the marina. In the morning they welcomed us by being really helpful. Paul lent us some local currency until we could find where to use our Visa card, and while I was at the market buying fresh food, he helped Ian get the electronic C Map functioning. Another couple lent us local cruising guides on CD, which we were also able to copy. More visitors included a local lady who recognised our boat. She had become friends with the previous owner when he had been in New Caledonia. We had gone from the peacefulness of the ocean to a flurry of social activity. Soon after we had received clearance, we needed to move out of the visitors' berth and into a normal berth. There were plenty of offers of help and soon we were tied up in our new berth.

After that, there were all sorts of practical things to do. I went off to find out where I could use my Visa card to withdraw cash. I discovered that the ATM inside one of the banks worked for foreign cards, while the ones outside didn't. Communications was my next priority, but it turned out to be much too expensive to have a temporary mobile phone. I was told that you could buy a phone card at the local bar to use in a public phone. We found that worked satisfactorily and were able to phone family in Australia, to let them know we had arrived. At this time, changes in technology were in such flux, we just had to adapt to local conditions.

We also had to work out where we could use the internet for email. We were anxious to hear from our nomadic elder son. He had crossed the Atlantic to Portugal and was intending to return to South America. When we discovered an internet café, there was no word from him and we were faced with a French keyboard. A week later we received Jamie's email from the Azores in the Atlantic off north-west Africa. He was returning.

After the spinnaker had blown out and it was in the cabin out of the wind, I was able to examine the damage before stowing it under the vee berth. I thought I might be able to mend it. To my delight I found that one of the other cruising boats in the marina, *Adios*, had a sewing machine on board and Ellie was happy for me to use it in their cockpit. I re-stitched the sail into all the outer tapes, a job I wouldn't have tackled without a machine. The repaired sail was going to be useful, but only in extremely light conditions or it would tear again. The head was weak.

While I was sewing, Ian was constructing a fold-up shelf for our laptop. We needed it securely available beside the chart table if we were going to use the electronic charts. With the assistance of some elastic cord, the computer stayed firmly in place whether the shelf was down or folded up and secured against the cabin side.

Finally, we were ready to start investigating the culture of this Pacific neighbour that is controlled by the French. That night was the eve of Bastille Day. Along with a group of other yachties, we headed downtown and queued for lanterns, then joined the lantern parade through the streets to the Place des Cocotiers, where an outdoor concert was scheduled to take place.

Fireworks started the entertainment, followed by different cultural groups each presenting an item, mostly dancing in different styles: hula dancing, native Kanaks in traditional paint and grass skirts, women in missionary style cover-up dresses and some Melanesian girls dancing very suggestive moves – as well as some modern dance. The performances culminated in a dragon dance by the Chinese community. We were well entertained and interested in the different cultures.

Ian and I noticed that despite numerous children running about, there was no screaming, no crying, no hyperactivity. People of different racial backgrounds seemed to mingle happily and there appeared to be little obesity, perhaps because the French culture doesn't embrace 'fast' food. We saw similarities to life in the French islands we'd visited in the Caribbean during the 1970s.

The population of New Caledonia in 2006 was about 220,000, of which 80,000 lived in Nouméa. Almost half the population of New Caledonia is Melanesian (Kanak), about a third European, and the rest Polynesian, Indian, Vietnamese and various other ethnicities. The Melanesians tend to live a more traditional lifestyle in rural villages and the other ethnic groups populate the urban areas.

The next morning, there was another parade with a far greater audience. The navy, army, military police, civilian police and fire brigade marched and there was plenty of brass band music. We were surprised there was even a contingent of Australians who marched representing the RSL. These two parades and the previous night's concert provided an enjoyable glimpse into the cultural life of the local people. We noticed there were fewer Kanaks in the audience lining the street for this parade.

Over the next two days, we prepared ourselves and the boat to leave the Port Moselle Marina. After paying our bill and returning the shower room key, we motored around to the next bay, where we anchored for the night.

Our plan was to visit the L'îlot Amédée. After Ian set the anchor in some clean sand near the island, we jumped overboard for a swim before lunch. A lot of the coral close to the boat was dead or only beginning to regenerate. Obviously, tourists had to stay close to their tour boat thus contributing to the coral depletion, but we were free to swim where we liked. By swimming further afield we found much nicer coral closer to the pass where the water was colder.

It was several years since we'd snorkelled in coral waters and it was wonderful to see very colourful parrot fish again, as well as numerous kinds of fish new to us – some bright lemon-yellow, some like army camouflage, some black with elegant iridescent blue tails, or edged with white.

On the sea floor, we noted several varieties of *bêche de mer* (sea cucumbers) and a few brightly coloured sea snakes. The sea snakes were thin, up to a metre long and vividly striped. They are venomous, though their small mouths make it difficult for them to bite adult humans. Bites are rare, because the snakes (or krait) are unusually placid for wild creatures. Even so, I preferred to avoid them.

When we landed the dinghy ashore that afternoon, we saw snake tracks leading from the water over the sand, though on our wanderings around the island we saw no snakes, thank goodness – we didn't even hear any slithering through the grass.

We were happy when we located the lighthouse keeper who opened the door for us to climb up the 45m iron-rung ladder to the top. The view seawards, north over the lagoon, was stunning. I love to step outside and walk around the top of lighthouse towers where the wind buffets but no salt spray can reach.

Ian was concerned about staying overnight in an exposed anchorage, so we temporarily returned to Nouméa anchoring in Baie de l'Orphelinat.

View from top of Amédée lighthouse

We motored out next morning for Baie de Prony in extremely calm conditions. Prony, lying on the southern end of Grande Terre, the main island, is a very large bay with two sections – North-West Baie and South-West Baie. We dropped anchor in North-West Baie and went ashore.

Just behind the beach was a sign: Propriété Privée. Chien Méchant. (Private Property. Dangerous Dog.) The French have very different ideas about the rights of people to have access between shore and land. The British do not allow anyone to own land beyond the high tide line. The French allow property owners to prevent access to the water in front of their property. I think we both were miffed about this when we wanted to have a walk ashore, so we ignored the sign. We climbed over the rickety barbed wire fence and walked through the deserted property, which gave the appearance of an abandoned resort. We followed a dirt road through the area, past some chooks (hens) and peacocks, ducked under the chain of a padlocked gate and continued uphill for twenty minutes.

We heard running water. A few metres from the road we could see a waterfall, spray and droplets sparkling in the sunshine; at its base, some inviting pools. We backtracked and found a path down to the pools, where we stripped and immersed ourselves for an enjoyable freshwater bathe. Freshly washed, we returned to *Osprey*. This had probably been the drinking water supply for the resort. As we didn't use soap or shampoo, we wouldn't have contaminated it too much. And we'd seen no sign of habitation, much less the 'chien méchant' during our walk.

A weather forecast informed us that the very calm conditions we were experiencing were to be followed overnight by a south-west trough with squally weather. We took *Osprey* across to the south-west corner of the bay and re-anchored in a more sheltered position.

The bay was still calm when we rose for an early breakfast on Thursday. Ian had decided to use the forecast wind to take us over to L'île de Pins (the Isle of Pines). Once we cleared the headland, with the genoa poled out, the wind fairly rolled us along. Our boisterous passage lasted nine hours. The wind in the squalls was gusting up to thirty knots and we needed to reef the sails when the wind backed. Inevitably, I was seasick again, spending most of the day in my bunk. As soon as we entered the tranquility of Kuto Bay, my nausea disappeared. We were very tired after all the motion and slept well that night.

On Friday we went exploring, and after partially walking a track to the highest point on this beautiful island, we returned to the main road. We walked some distance before we found a small *magasin alimentaire* (grocery store) where we were told that the stores boat came in only every second Saturday. Next morning, Saturday, we discovered we'd missed the early morning market which was held from 6.00 – 7.00am. We hadn't gone ashore until mid-morning, by which time the sun was shining, and the water a sparkling turquoise.

There were quite a few people out and about on bicycles. Ian became interested, suggesting we should hire bikes for a day. After a few false starts, we found a *gîte* (lodge) which would hire two bikes to us on Monday morning.

All Sunday and overnight, there were rain squalls. Despite a couple of rain showers early on Monday morning, we picked up our bikes at 8.00am. Ian was able to raise the seat on his, but mine was too rusty. It was the only women's bike they had available, so I rode it anyway. I think we covered about forty kilometres that day, visiting a local village, watching local boats (pirogues) racing and wandering along a deserted beach, where we found and used a public shower.

The disused prison interested us and we were able to walk through the historic buildings. Everything was so different to Australia. I think I would have been more comfortable in one of these cells, despite the chains on the walls, than in one on Sarah Island in Macquarie Harbour.

The old French Prison, Île des Pins

By the time we returned those bikes my muscles were aching and my backside extremely sore from the hard seat. Ian rode further than me – he returned to Vao, to the grocery store we'd found in the morning, and came back with the vegies I'd left there to be picked up later. We were tired but happy with our exploration of Île de Pins. Our most strenuous activity the following day was to massage each other's aching muscles.

A couple of days later, we left Kuto Bay, stopping at Bonne Anse in Prony, before returning to Nouméa the next morning. Having missed the markets on Île de Pins, I was keen to stock up on fresh food and also to do our laundry on a good drying day. There was no laundry room at the marina, but no-one minded that we hung the clothes out on a line around the rigging. I washed on the dock beside the fresh water tap, using buckets, and wringing everything by hand. Most people staying at the marina did it that way.

We caught a bus to the Marie Thibault Cultural Centre on Sunday morning. There was plenty to see and learn about New Caledonia and its culture.

Jean Marie Thibault Cultural Centre

One style of traditional native house

The centre is a fascinating place with its mixture of gardens, art gallery, museum and history of native culture. In particular, the surrounding gardens were spectacular. I was most interested in the full-sized samples of the different kinds of native houses from the time before European settlement.

⚓

We had been in New Caledonia for nearly a month, and realised that we just didn't have time that year to sail to Vanuatu as well before the next cyclone season. We found New Caledonia enchanting; wanting to see more, we hired a small car for two days.

Lake Yaté water supply for southern towns

New Caledonia has three Provinces: Îles Loyaute, Nord, and Sud. These are in turn divided into thirty-three municipalities. Apart from Nouméa and its surrounds, there are some small towns and the rest of the land is sparsely populated by Kanaks living in traditional villages. The majority of the population lives in the Sud (south) province.

We drove across the south of the island on the first day in our hired Getz, our objective being to visit the Blue River National Park. The Blue and White Rivers join and the French dammed them to form the artificial Yaté Lake, a permanent water supply for Nouméa and the other southern towns. Scrubby bush grows on the surrounding land.

We also found a picnic ground to eat our lunch, with a nearby walk. Many of the native trees were labelled, including sixteen varieties of palm trees and a giant houp tree, a threatened species of rainforest tree native to New Caledonia.

The next day, we drove in a different direction north and east, through Boulouparis to Thio, where we spent a couple of hours at the mining museum. There are several big mines in the south of Grande Terre, mostly for nickel, but also chromium and cobalt. It is sad to see great gashes in the landscape where mining operations have taken place and there appears to be little rehabilitation of the land after the mining is finished.

Returning through Boulouparis, we visited the essential oil distillery there. The local oil is Niouli, from a tree related to the Australian tea tree. Winter was the wrong season for the distillery to operate, but I was able to buy some oils later at the market for presents to friends at home. The niouli essential oil is made in spring and early summer from the leaves, shoots and twigs of the trees, which are steamed until the oil is extracted

There was so much to see and do in this island country. After discussions with many people we decided to sail up the west coast of Grande Terre and return via the east coast. I loved the idea of sailing inside the lagoon where I would be able to indulge my pleasure of snorkelling over coral. To do this we needed more charts, borrowing and photocopying those lent to us by an American couple on *Tyche*.

Amidst all the preparatory chores for our circumnavigation of Grand Terre, we rang family members. David had announced via email that he had met a girl called Heather. He'd never made any announcement previously when he met someone new, so we knew this girl was important to him. Now I was keen to meet her – another reason to put off going to Vanuatu – but we did have time to circumnavigate Grande Terre before heading home.

9 Circumnavigation of Grande Terre

Fig 3. Map of New Caledonia *

Grande Terre is the largest of the New Caledonian islands. Following advice from a local yachtie, we planned to sail up the west coast first. We paid our bill at the marina and were ready to leave in the early afternoon; the engine was running and we'd just cast off our mooring lines when Jim Cate appeared. We had last seen the American couple from *Insatiable II* in Tasmania the previous year. We tied up again and turned off the motor.

Jim and Ann had just arrived from Australia. We visited their boat after the customs officer left and stayed chatting with our friends until the immigration man arrived. By that time, it was too late to go far, so we pulled out of the marina and anchored in the harbour for the night.

Our departure was delayed further because it was wet overnight and into the morning. We always put up with rain when it starts en route, but we were reluctant to set off while it was still raining. We cleared the harbour at 11.00am when the showers ceased, and tacked north as far as Îlot To Dho, so pleased to be out in the lovely lagoon again.

* Map courtesy of Ward, Alan, 1935: *Land and Politics in New Caledonia.*

The next morning we moved on to Île Moro, where we stopped for lunch and then went snorkelling.

In the lagoon we towed the dinghy behind *Osprey*, making it quicker and easier to get into the water when wearing our wetsuits and fins. After the initial chill of the water, it warmed inside my wetsuit and I felt a sensuous luxury about swimming. The lagoon was relatively shallow, and felt tropically warm on my face and hands. There were a variety of corals and dozens of brightly-coloured fish: bright yellow ones, blues, greens, red and black with white – many kinds we'd never seen before. What a wonderland! I drifted with the tide, gazing at the natural glory before me.

How wonderful to find the coral here had not been bleached as large patches of it had on the Great Barrier Reef when we snorkelled north of the Whitsundays in 2000. As climate change has gained pace, but I know that huge areas of the Great Barrier Reef off Far North Queensland, have been seriously bleached and damaged over multiple summers.

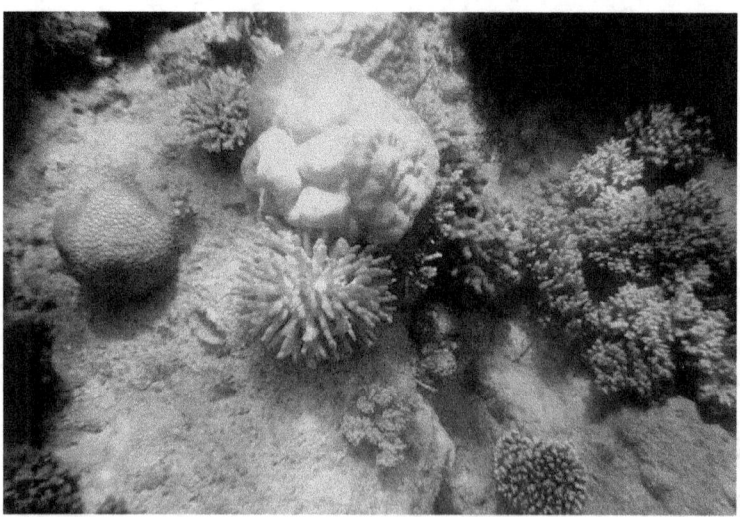

Reef corals

We found a rocky area with oysters just as the tide was going out. Ian returned to *Osprey* for hammer, screwdriver, leather gloves and a small bucket – our kit for prising oysters off the rocks. Although the shells were quite big, the oysters were very small and obviously young, so we took only about a dozen – just a taste. Seafood is so much tastier when it is freshly harvested.

The next morning, Ian woke up feeling unwell. That, combined with strong southerly wind squalls, convinced us to stay put for a couple of days. He rested while I did chores and read. That afternoon, I went snorkelling again. I was amazed that my love of underwater gardens took away all fear of being out there on my own, and all fear of sharks.

The following day, I received an unwelcome surprise. I was in the galley preparing breakfast when a blue and grey striped sea snake slithered out of the starboard quarter berth. I squealed and jumped into the cockpit. Ian was on deck.

'Ian! Help! There's a snake in the galley. It's gone under the stove burners.'

He went below and found the snake, grabbing it behind its head. I wanted him to throw it overboard immediately, but he wouldn't.

'Not until you take a photo of me holding it', he insisted. I did so, but from inside the cabin, which gave an odd angle to the picture.

The sea snake Ian caught in the galley

I was pleased to see it swim away. It certainly hadn't been there a couple of days earlier when I checked over the fruit and vegies for items that needed using first. Could a snake come up the anchor chain? Apparently so as I now know they are adept at climbing.

Mid-morning, we were ready to leave when I opened the engine cover to turn on the water cooling. Assuming I'd turned it off previously, and without thinking carefully, I turned the lever for the valve. Ian noticed the engine overheating and yelled at me. *Mea culpa*. I realised I must have turned it off. Fortunately, the exhaust-box hadn't melted, so I was forgiven. This was an on-going problem. Ian often forgot that I considered getting the engine ready to be my job, and I failed to learn which way the lever went for 'on'. Why didn't we ever mark the lever to show which position was off and which on? Or just make it Ian's job?

Our next stop was an island near the Passe de St Vincent – just an over-nighter. We had intended to go through the pass the next day, but when we downloaded a weather fax, we saw that there was likely to be a moderate westerly almost on the nose for the next couple of days. We reconsidered and continued inside the lagoon the short distance to Île Isié, where we swam ashore because the dinghy was already rolled up in the forecabin.

The isolated, white-sand expanse was dotted with lots of shells from pipis, mussels, cockles and conch, and we walked the length of the beach. We dug into the sand at the water's edge with our toes looking for live pipis, but had no luck. Fresh shellfish free for lunch or not, this place was paradise, although we certainly weren't Adam and Eve. I could see the allure of owning one's own island.

That afternoon, Ian was doing some engine maintenance when he discovered that there was a hole in the exhaust water-injection bend. We didn't have a spare on board and it seemed unlikely we could acquire a replacement easily. Ian, ever resourceful, improvised a repair by cutting a piece of aluminium from the lid of a sardine can and then, using silicone sealant, covered the hole. He used hose clamps to hold it all in place until the sealant dried. The alternator belt was also giving us some trouble. Ian tightened it again. He started the engine early next morning to check that all was well. The silicone didn't overheat so instead of returning to Nouméa, we were able to continue our circumnavigation of the island as planned. We were pleased this had happened before we left the lagoon. Neither of us wanted to curtail our adventure.

Ian cleared away his tools and I locked down all the cupboards before we headed through the Passe Isié into open waters. Outside the reef, I was nauseous again, but this time the anti-nausea Relief Band kept my seasickness under control. It was an overnight sail up to Koumac, where we entered the pass back into the lagoon.

On the way in, we motored by a tiny island just like the ones in cartoons – small, round and sandy, with a palm tree or two in the middle. Irresistible! We turned around and anchored off, donned wet suits and swam ashore. It was magical to walk around this tiny coral *îlot*. We met a French couple there, who were on holiday and with whom we chatted briefly in our hesitant French.

I served lunch while we motored to Koumac's outer harbour. There, we unpacked the dinghy and Ian rowed ashore to check out the tiny marina. He booked a berth for the next day.

The marina lady offered us a lift to the village at 10.00 the next morning, Monday, which Ian accepted. We bought a couple of frozen chickens which had come all the way from Brazil, and some vegetables to top up our supplies. At the butchery, we ordered meat for the next day.

A two-palm tree islet in the western lagoon near Koumac

Koumac is a small town in a rural area – mostly small dairy farms with black and white cows contentedly chewing in the lush green paddocks. It reminded me of the small dairy farm my parents owned when I was a young teen. We walked back along the unsealed road on Tuesday morning to collect our meat and buy some fresh fruit. Ian carried most of it in his biggest backpack.

We wrote letters to friends without email, and we used the internet at the marina to catch up on email to family and send out a newsletter. Ian ordered a new injection-bend for the exhaust to be sent to Port Moselle as we didn't want to rely on the repair for the remainder of our voyage.

Both days we were at the marina, the wind came up early, gusting onto the land. In preparation for leaving, we turned *Osprey* so that her nose was pointing out of the berth. Although the Brolga reverses under the engine with more control than *Realitas* did, it is still dangerous to be caught by the wind when backing. A lot of expensive damage could occur if the boat crashed into a pylon or another vessel. Injury to ourselves was also likely when trying to fend off, as we'd found when we tried backing *Libelle* out of her berth in Marmong Point Marina in 2001 when we'd been trying to sell her.

When we went to pay our marina bill, the marina lady charged us 800 Pacific Francs for driving us into the village on Monday. That seemed cheeky, because she'd said she was going anyway, but it was too hard to argue with her in French, so we paid.

From Koumac, we were able to continue within the lagoon. The wind came in behind us and we had a glorious sail up to Île Tanlé. The isle has a small hill 142m high on the southern side. Ian went ashore to climb it while I rested my ankle, which I'd sprained just before departing Koumac.

'The hillside is covered in weather-worn rocks and gravel, and there was head-high scrub. Even more surprising was how little wildlife I saw – just a few ants, a butterfly and three spiders. I was disappointed,' Ian said on his return.

My ankle being too sore to climb up the boarding ladder, I used the dinghy as a landing platform and went for a swim. That way, I was able to have some exercise too.

The next day we saw the only other yacht in the lagoon since we'd turned north from Nouméa. It was one of the Australian yachts we'd met in both Coffs Harbour and again at the marina in Nouméa. We were really surprised at how few visiting yachts were out exploring the lagoon.

A few more hours of sailing brought us to the Baie de Banaré and the village of Poum, the northern-most village on Grande Terre. The guide book told us it was once a thriving copper mining town, but that industry had ceased in 1938.

The area gave me the impression of being a tourist destination minus the tourists. The scenery was great, there were numerous offshore islands and the market area was obviously vibrant when in use. No doubt it came alive when cruise ships visited. We saw no sign of other tourists, yet there's a hotel well hidden among the trees near the shore. Nor did our guide book mention that the lagoon off Poum is part of a World Heritage area.

Landing on the beach, we walked until we found the local *magasin*. The shop was run by a Thai woman and her Australian husband.

'There's no market on Saturday morning,' said the lady, 'but there's a Kava Bar open tonight at 8.00pm.'

It sounded intriguing and I wouldn't have minded watching for an hour or two, but not enough for us to go at night. Trying to find our way back on board *Osprey* after dark was not appealing. And much as I would have liked to try kava, I knew that with our extensive food sensitivities, kava was not for us.

We walked down the inland road, but all we found was a bus stop with bright murals on the shelter. Along the other road, there was a locked church. I wondered if it was locked against the locals or the tourists.

One small offshore island (Mouac) had seating near the beach and jetty, with tables and palm-frond umbrellas. I could envisage tourists sitting there, sipping cocktails. Not far from the beach on Mouac was an area of coral labelled 'Underwater Garden'.

I had to investigate and yes, it was an attractive site. There was little sign of damage to the coral such as we'd found when sailing *Realitas* in the Whitsundays. Again, my estimation was that it was not heavily visited, because the coral and fish were superb – the best we'd seen so far in New Caledonia. They were similar coral species to those we'd seen elsewhere in the lagoon, but in excellent condition. Ignoring the other islands to the north, which appeared uninhabited, we moved off around the point to the eastern side of Grande Terre.

There was no wind and we motored for five hours to reach the Baie de Pam. This had once been a chromium mining area, long since abandoned and still mostly undeveloped. Any scars from mining had overgrown. Although we enjoyed our walk ashore, we met no one except a couple of French tourists in a car and saw nothing of great interest.

Moving further down the east coast, we discovered there were quite large discrepancies in the digital CMap chart. Ian had been checking the CMap against the paper charts all the way up the west coast as far as Poum, and it was mostly accurate because that's where tourist and commercial ships plied. I found it very disquieting now to see the computer screen showing us stopped well inland; at times it showed us to be up to half a kilometre over reef or shore. Often the CMap made no sense at all. Vigilance was necessary except in areas where ships brought tourists and where they came in to load ore. Those areas were shown correctly on the charts. As usual, money talks.

Off Ballade, there is an Îlot surrounded by a large area of reef. We were intrigued by the mention of a monument on this islet, so we anchored off and went ashore. The whole area was of historical significance. Captain Cook landed here in September 1774; a French expedition arrived in 1779; and the monument (leaning) was for Captain Huon de Kermadec, the first European to die in these islands.

Monument to Kermadec

A French mission had been set up at Ballade, but it wasn't until 1853, right here, that the land of New Caledonia was officially declared French, well after the chromium mine at Pam had already closed.

Four local young men were cooking fresh, reef-caught fish for their lunch on a small fire on the sand. They greeted us in a very friendly manner and invited us ashore to their village. One came on board *Osprey* to show us the best route through the coral heads to a safe anchorage, and then the others picked us up in their new aluminium dinghy and took us ashore with them.

We asked where we could buy fresh produce and one of the young men asked his mother to sell us vegetables. She dug up fresh manioc and yams and cut a lettuce for us from her garden.

The guys then returned us to *Osprey*. They had to work at one of the nickel mines the next day.

There was a fresh water tap by the shore. Next morning, we brought our laundry ashore and left it to soak while we went walking to the old mission. This church wasn't locked and the entrance was decorated with shells, while an enormous clam shell formed the holy water font at the back of the church.

Ballade Church font made from a clamshell

On a noticeboard we saw a newsletter to parents written earlier in the year. It was the age-old request for parent volunteers at the school, but it also berated them for their children's poor attendance. About sixty percent of New Caledonia's population is Roman Catholic.

Later that day, we arrived in Pouebo. Our cruising guide informed us that there were a shop and petrol station next to the bridge over the river. We motored our dinghy up the river to the bridge but there was no sign of petrol bowsers or shop. Having little success with asking children coming out from the local school, we asked the driver of the school bus when it pulled up.

He smiled broadly. 'Come with me. Climb in. I will take you.'

To the delight of the young school children, we climbed aboard with our empty petrol cans. The two sitting closest to us were very eager to practice their English and to tell us about their classes. The bus driver dropped us off at the 'servo', promising to return for us when he had delivered the remaining children home. The two boys we'd been talking to grinned widely as they handed our containers down to us.

About fifteen minutes later, the driver returned as promised and dropped us and our full fuel cans off at our dinghy. This was the second time in two days that the local people had been exceptionally kind and helpful to us.

Before sunrise, we moved on towards Hienghène, arriving there in the middle of the day. This eastern side of Grande Terre had large stretches of low lying reef and fewer islands than the west coast where the height of tide had not been of great concern. We needed to watch the tide and avoid the sun in our eyes so that we could see the coral reef clearly while navigating and anchoring.

Hienghène is a delightfully thriving modern town where, at the time of our visit, the old town centre had been recently rebuilt for tourism, but much also catered to the locals – a new high school, modern sports centre, public telephones, a cultural centre, library and hospital. The streets were well paved, including footpaths. The villages we'd seen in the north had unpaved roads and no footpaths.

Obviously, the town was being 'groomed' as a showpiece for tourists – or possibly to win an election?

The small marina was also new, sporting fresh water and electricity for yachts. Unfortunately, the fuel supply was still in the process of being installed. As we had motored quite frequently when there was little wind, we had hoped to get another thirty litres of diesel.

Everything in town closed by half-past four and the sun set behind the mountains by 5.00pm. We returned to town early next morning. The cultural centre wasn't open yet, so we decided to walk to the look-out. En route, some locals picked us up in their car and drove us there. Carelessly, I left my SLR digital camera on the back seat of their car, a fact we noticed as soon as they were gone and I wanted to take some photos. Had we lost it for good?

Back in town, we went to the Information Centre, where I explained our loss and described the car and men. While we availed ourselves of the modern communications in the centre to check email, the lady in charge tracked down the car driver – the wonder of small towns where everyone knows everyone else. The car owner promised to return the camera early that afternoon and was as good as his word.

We set off for the Cultural Centre, which was open by then, and thoroughly enjoyed the displays of native culture, the museum and an historical photography exhibition. Everything was beautifully laid out amid well-kept lawns and gardens.

Later, as I was walking in the town, a car stopped; it was our driver from early in the morning. He was anxious to know if we'd retrieved our camera after he'd left it at the Information Centre.

'I didn't know you'd left it behind until I received the phone call. I'm happy you have it back.'

'We're happy too. Thank you for your trouble.'

That afternoon we walked along the beach to a famous cave to photograph the island just offshore, which was lit up in the last rays of sunshine. Hienghène is noted for its diving, but I have no memory now why we didn't go snorkelling there – possibly just because it was a tourist site.

We moved on again after two nights, our destination the Port de Touho. In the afternoon, we went ashore to the village and found an agricultural festival in progress. There was only a small amount of fruit and vegetables for sale because the local produce market took place on Fridays. Other stalls sold clothing, hats and bric-à-brac. I suppose I had been expecting local food markets to be on Saturday mornings as in Australia, but market day varied from village to village.

La Poule – the Hen, outside Hienghène

The variety of fruit and vegetables available at these small towns was limited, which is why we checked out what was available at every market. There was always something we needed.

After checking out the festival and finding nothing of great interest to us, we went in search of diesel and found that the store was open from 2.00pm until late afternoon. With the addition of another thirty litres our fuel tank was full, giving us the satisfied feeling that maybe we had enough now to get us back to Nouméa. We were also able to top up our drinking water from the tap on the wharf. Such mundane chores took up a lot of our time and energy, as did walking everywhere on shore.

Touho held none of the natural beauty of Hienghène, and we left again early on Sunday morning. It was pleasant to have a good sailing wind instead of the noise and vibration of the motor. We stopped near a sandy *îlot* and, because the day was warm, after lunch we swam ashore past the coral heads wearing our stinger suits, for protection against venomous jellyfish, fins and masks. Ian wore his mask for our walk around the island, because it has lenses for his poor vision and he hadn't carried his spectacles ashore. We saw a group of children with an adult there and we were amused when the children screamed and ran away after sighting Ian. We must have looked very weird to them.

At Ugué Bay, there was no settlement – just a nickel mine and a loading wharf. Presumably, the airport at Touho was to service the mine.

The wind and swell funnelled into our anchorage, making for a very rolly night and we were pleased to pull up the anchor when daylight returned.

It was 25nm to Kouaoua, and we were disappointed to note the large scars from mining on the hillsides as we sailed south. I thought the mining companies should have to do more to rehabilitate the landscape.

However, another friendly local made us feel better. After we'd found some frozen supplies at the grocery store, he drove us back to the boat ramp, waited while we dinghied to *Osprey* to collect our fuel cans, and then drove Ian three kilometres to a fuel station for diesel, before taking us back again by dark.

Such kindness was widespread among the locals, both Kanak and those of French origin – yet in Nouméa, we'd been told by other visiting yachties that the indigènes were unfriendly.

The need to buy diesel was ongoing. We had not expected there to be so little wind and to have to motor so much. Every time we thought we surely had enough fuel to get us back to Nouméa, we ended up using the engine all day.

In the morning, we attended to chores – phone calls to Australia from the box on shore and we made good use of the fresh water tap near the boat ramp by doing laundry, showering as best we could while wearing clothes, and hair washing. Neither of us was feeling very energetic, so we called it a rest day, hanging out on the boat, sleeping and reading.

There was a very pleasant bay six nautical miles further on. As usual, I trolled a fishing line, but I forgot to take it in while we tried to find a space to anchor away from coral heads. That was where I caught my only fish of the entire voyage – a coral trout. I dislike killing, cleaning and filleting fish, and I only kill for food, but how nice to have a fresh fish dinner. That particular fish was delicious.

We went swimming in the bay, only to find most of the shallow coral damaged. Six or seven metres down, it looked healthier. I never tired of looking at the shapes and colours of the coral and the brightly-coloured fish living there, but I hated to see it damaged by swimmers and boat anchors.

Jan with the coral trout she caught

All this time, we'd seen only four other yachts – the Australian vessel under sail near the north-western tip of the island and now on the east coast, a local vessel in Hienghène and a couple of cruising catamarans at Touho. However, as we proceeded further south, we began to see more foreign yachts. The most we'd seen in any one anchorage other than in Nouméa was at Île de Pins.

Our next overnight anchorage was at Canala, a sheltered place with good anchor-holding to ride out a forecast gale. We went ashore to the local village, where we were able to top up our food supplies again. We visited the grocery store at each village, usually finding some item we'd not been able to get previously. Strong winds and rough sea buffeted us, so we stayed at anchor that day.

By the time we'd arrived in Canala, it was the first of September and we'd been circumnavigating Grande Terre for just over a month. Our visa permitted us to bring our boat into the country with us for a total of ninety days, and since we had arrived a week into July, we had to leave by the first week of October. Time was running out.

A couple of days later, at Lavaissière, four young Kanak men visited us in their tinny, powered by a 40hp Honda outboard. We were a little surprised at this level of spending, but it turned out Glenne and his cousin Lionel were shift workers at the Nakéti nickel mine, while Rudy and Richard worked at the port loading ore onto ships bound for Townsville. New Caledonia has nearly half the world's nickel deposits and exports about twelve percent of global production.

Rudy told us his family owned land in the local village. He worked two five-day shifts per month, spending the rest of the time raising pigs. The dinghy was his. He sold excess pork to the supermarket in Canala. He also fished and helped his wife with their vegetable garden to feed themselves and their six-year-old son. This seemed to be a fairly typical existence for the more well-off locals, blending a traditional lifestyle with some paid work.

Ian with four young Kanak men from Lavaissière

The guys had beer and wine with them. When they came on board *Osprey* they talked volubly, expressing fervour for independence from France and even revolution. The Kanaks hate the discriminatory colonial policies they feel have been levied against them throughout the period of French control.

Despite visiting museums and cultural exhibits, we had been unaware that native land is communally owned. There are approximately three hundred different land-owning clans in New Caledonia. The land is not sold, but is sometimes leased.

The native peoples mostly feed themselves by fishing and growing all their own vegetables and coconuts. We observed many women gathering shellfish on the shallow reefs of the east coast of Grande Terre. Those whose family members work in the mines supplement their traditional fare with store-bought food.

I suspect now that the water supplied to the taps near the shore of each native village is also communally owned. In Australasia, we are so accustomed to the local council supplying water to a public tap that it didn't occur to me that we should probably have requested permission to use the village water.

Rudy and his friends invited us to accompany them ashore where they showed us their pigs which were fed on coconuts, and their garden. They had solar panels for electricity and water tanks. Some of the larger villages had electric power from the grid (including at Canala and Touho). There was a small hydroelectric generating scheme on the Yaté River which supplied the towns in the south.

Rudy gave us some cassava and told us it needed to be stored in the fridge if we weren't going to cook it straight away. Some of the cassava we'd bought at Ballade had already rotted.

Later, when the young men had gone off to work, Ian took our dinghy ashore and filled several of our four-litre water bottles so we could top up the water tank.

We sailed on to Thio next day, arriving about 9.30am. We had been there earlier by car, so didn't stay long. We did go ashore for a walk and a friendly local gave Ian a lift to the fuel station – another opportunity to top up our diesel tank, since we continued to use the motor every day. There was still little sign of the current Pierre had told us of.

We left the lagoon and sailed outside to re-enter at Port Bouquet on the southern section of the east coast of Grande Terre. There, we anchored off Île Nemou where we went snorkelling. The corals were different and more colourful than in the north or on the west coast While there were fewer fish, there were some varieties we'd not seen further north. In particular, we were intrigued by some needlefish about 70mm long, clear and with a black stripe of backbone showing. The bodies gave the appearance of being feathered. They were keeping station vertically over their coral home. There were also some red starfish and lots of sea cucumbers of several varieties from small black blobs to many that looked like Vienna loaves of bread. I spent so long in the water that, despite my wetsuit, I was quite cold when I emerged.

We had initially intended to visit the Loyalty Islands to the east of Grande Terre, but when the south-east trade wind reasserted itself, we decided it was too late to go to the Loyalty Islands. We had only three weeks left on our visa and we would have to sail hard on the wind to get there. We thought we might go on a later trip.

After our swim, we moved our anchorage to Toupiti, at the west end of Nemou. This entailed crossing an uncharted region. Even though it was low tide we found no place less than 15m. There were a few coral heads, but inshore of those we were able to anchor in sand and mud clear of coral. Anchor chains cause most of the severe damage to coral, but tourists touching, or even standing up on it is also detrimental

We went ashore at the head of the bay after lunch, where we found a house and an extensive vegetable garden. There was no one home while we were anchored there so we assumed they were away at work. Fishing buoys and nets were hung in the trees. I took our laundry ashore and washed it at the outdoor tap next to the beach. It was a day with exceptionally low humidity, so I sanded one side of the dinghy floorboards and varnished them, adding grit as anti-slip.

Next day, I swam around the coral heads where I saw some very nice corals and up to eight different stag corals, some a vibrant blue. There were many fish too. After swimming, I went ashore to rinse off in fresh water and wash my hair. Apart from using some fresh water, we didn't touch anything in the garden or around the house. Strictly speaking, I suppose, we were trespassing and, now I realise, even stealing their rainwater.

While we were at that peaceful anchorage we watched a flock of New Caledonian crows which, we knew, are regarded as among the smartest birds in the world because of their toolmaking. Many experiments have shown them capable of complex problem-solving. In one experiment I have read about, the crow was presented with some tall clear glass cylinders holding varying levels of water. Each jar had a meaty treat floating on top where the crow couldn't reach it. Nearby, the experimenter left a pile of stones. The crow took the stones and dropped them into the jar where the water level was highest, thus bringing up the water level until it could reach the treat. It repeated this with the other jars until it had reached all the treats. I notice time and again that we humans far underestimate the intelligence and abilities of other creatures.

From Toupiti, we moved on south, motoring again until late morning when there was enough wind to sail. After three hours, the wind dropped off, and then we motored for another hour and a half before finding a safe anchorage. We continued to Yaté in the morning. It was Sunday. There, we walked to the village but didn't find the 'magasin' or any shops.

Children we met seemed unable to understand our bad French. There are over thirty Melanesian dialects in New Caledonia and we didn't try to use any native expressions. We should probably have learned the basics of 'hello' and thank you. Too late, I've realised that people like to be greeted in their own language.

Eventually, we came across a friendly French/New Caledonian couple, who explained that no shops opened on Sunday. They insisted we take two pieces of venison from their freezer and some apples. Someone had told us that New Caledonians were unfriendly, but we experienced generosity from both Europeans and the native peoples – the Kanaks.

Three boys we'd spoken to earlier were waiting on the beach for our return. One asked in very polite French if they might come out to look at our boat. We agreed and they all crammed into the inflatable dinghy with us, Ian being the only one to get a wet bottom.

All three appeared about 11 or 12 years old. Two, who were cousins, seemed academically bright, but the smaller boy, who could hardly write his name, understood immediately how our windvane self-steering system worked. We thought he might be the son of a fisherman. I gave them chocolate and pieces of ginger as snacks and water to drink. We took photos of them and promised to send them copies. We had also done that with the other young men.

We heard that there was a waterfall at Goro and decided we'd like to climb up to it. To get to Goro, we had to exit the reef and re-enter at Havillah Pass. Ashore, we soon we heard splashing, shouts and laughter.

A group of young people were swimming in a pool downstream from the waterfall. We asked for directions to the pathway, but two young teen girls decided to take us.

Marie Joe and Kelsey, who led us to the waterfall at Goro

We were amused when they asked if we could manage because it was steep. We managed! The view from the top was wonderful. Like all the young people we met, these girls were incredibly polite, careful and considerate. During the climb they frequently proffered me a hand if they thought I might slip. We sent photos of the girls to them.

View from top of waterfall – Osprey *at anchor*

The girls directed us to the snack bar, where they said we could buy yams. We found the couple who ran the snack bar, bought eggs and yams and chatted to them. I was very impressed with myself. After all this time in New Caledonia, my oral French was improving. I managed to crack a joke in French and the couple seemed to understand – or at least they laughed!

When we moved on the next morning, we turned west – we had now travelled the entire east coast. Our next destination was Port de Boise (Woody Bay). Just over an hour later, we dropped anchor.

The weather turned dull at Woody Bay

There was a stream at the top of the bay and alongside, a water pipe maintenance track. On the track, we passed an older French couple (grandparents?) and three Melanesian boys, who were walking extremely slowly. After reaching the top of the track, we returned to find the group only 300m further on. This might have been of no consequence, but we all arrived at the stream together, which meant Ian and I couldn't strip off to bathe.

The sky was overcast, making it unattractive to swim or snorkel in the bay. Back at the dinghy, we motored to another stream further away, where we washed ourselves. Whenever possible, we preferred to bathe in fresh water. Our clothes and sheets became slightly damp from the salt off our bodies whenever we washed in sea water.

After staying in Boise overnight, we moved on to Prony next morning. We were now in territory we had already visited earlier so, in effect, we had completed our circumnavigation of Grande Terre.

Cold squalls and gusty winds met us in Baie de Prony. Ian let out the genoa and we raced along. At the far end of that very wide bay, we found shelter to anchor and then went ashore to stretch our legs. The chart showed a warm spring in the vicinity, which we were keen to find. We were fortunate to meet up with a local family who showed us the way. They were also on a yacht and had laundry to do and freshwater to collect. We walked together to the cascade. The woman was Australian, Her husband a Frenchman, a recently retired headmaster, so communication was excellent.

The water in the concrete pool was just warm enough at 26°C and soaking there proved a pleasant experience on an overcast, cool day.

The weather became unpleasant again that afternoon and overnight with strong winds and rain. Just as we were deciding to leave the bay, there was a break in the clouds. We changed our minds about going, because we had been hoping for sunshine to view l'Aiguille (The Needle). We changed into our wetsuits to swim at this renowned underwater pyramid formed by the upwelling of a hot mineral (sulphur) spring. It was marked by an isolated danger mark on the surface.

Coral has colonised the limestone needle and striking, black batfish with a white streak on the 'wing' live in its vicinity. Because of the sulphur content, many of the corals are yellow. I was amazed to see yellow branching staghorn – a coral which is normally various shades of blue. Some yellow corals looked like paper flowers, some large white ones looked like overblown roses. There were some little yellow growths in pairs, like two thumbs, and some single ones with a nipple on the tip. We felt very fortunate to have the opportunity to see this unusual and relatively rare phenomenon.

Feeling a great sense of satisfaction with ourselves for achieving an exploration of the island in the manner we prefer, in mid-September, we sailed *Osprey* back into the marina in Nouméa.

I enjoyed being able to visit the early morning markets again. It was also reassuring to catch up with family on email. We had received conflicting accounts about where Jamie was, but it turned out he'd sailed to Portugal, where he caught up with his friend, Sylvain, from their days at the University of Zululand. Jamie wrote to us from the Azores en route across the Atlantic back to Brazil and Argentina.

David was with a friend on her Top Hat yacht in the Brisbane River, and his new girlfriend, Heather, was still in the picture.

Dot Vidgen and her husband, Vic, members of our CCCA yacht club, had come into the marina on their boat. Dot invited me to accompany her to the zoo and botanical gardens. She was great company and I had a really enjoyable time with her.

The bird exhibits at the zoo were magnificent. They even had a cagou, the national bird, which is rare. According to Wikipedia, 'it is the only surviving member of the genus Rhynochetos and the family Rhynochetidae …'

Cagou – the endangered national bird of New Caledonia

There were several kinds of peacocks as well as many birds I'd never seen before. Being tropical varieties, most of them were very colourful and beautiful.

Afterwards, we couldn't find a bus back to town (probably because it was Sunday afternoon) and we ended up walking several kilometres back to the marina.

On Saturday 23rd September, the celebrations for Caledonia Day were held. We went along, as did many Kanaks. However, unlike the other cultural events we'd attended, this one appeared to consist mainly of political speeches – somewhat boring for us

Between these busy social activities, we actually managed to prepare for our voyage back to Australia and several days later we cleared customs. We were partly sad to be leaving New Caledonia, yet pleased that we'd seen more of this beautiful group of islands than originally planned. Somehow, we thought, we'd get back again and go on to see Vanuatu, but more than ten years later, this has not eventuated. Health considerations and increasing age have ways of diverting good intentions.

We cleared Port Moselle marina late morning and were out of the lagoon and through the Pass du Nord two hours later. The wind was aft of the beam and we ran south-west at seven and a half knots most of the night, making excellent progress in the right direction. The water swished past the boat and, lying in my bunk, I could hear it through the hull and feel the regular motion of *Osprey* cresting the swells and sinking into the troughs – not that the swell was big, just a peaceful, long and regular rolling which lulled me to sleep.

Over the next two days, I caught three fish, but failed to bring any of them on board. They must have been big, because they bent my hooks before getting away.

'I thought you were going to make up for all that money you spent on fishing gear,' said Ian, as I brought in an empty line for the third time.

'I'm disappointed too.' I had spent about a hundred dollars on line and lures for this trip and we'd caught only one fish on the east coast of Grande Terre.

We experienced good south-east trade wind sailing, with the wind just aft of the port (left side) beam. Occasional squalls followed calms and we used the engine at these times to assist with battery charging. In port, we used the flexible solar panels we'd bought second-hand two years earlier in Scarborough, but these were stored inside when we were at sea. The two glass solar panels on the arch and the wind generator were our other means of generating power. We used lights at night and also ran the freezer, computer, GPS and radio all the time at sea, so we needed a good supply of electricity – so different from when we sailed around the world in our Top Hat, *Caprice,* during the 1970s. Then, we used the battery only for engine starting and lights. There is a great deal of satisfaction to be gained from being self-sufficient.

We made excellent progress towards the NSW coast. By the evening of our fourth day, we were only sixty nautical miles north of our GPS waypoint to clear Middleton Reef. I was on the dawn watch again – my favourite time. I wrote in my diary: *Brilliant sailing – just how I like it. Gentle trade wind seas and enough wind on the beam to carry us along at 5-6 knots. A beautiful starry sky and a quarter moon.*

This was followed by light and variable conditions as we passed by the seamounts in the Tasman. When we were no longer in the trade wind zone, we picked up the temperate zone wind patterns and flew along with the sails winched in tightly for a north-westerly wind. We needed to keep upwind countering the southerly current as we approached the coast.

That day, Ian and I had a row. These are so rare that it was noteworthy. The wind had gone light.

'I think I'd like to put up the spinnaker,' said Ian.

'Not a good idea. It's too weak. It needs reinforcing at the head.'

'Why did you say the spinnaker was ready to use again when you haven't strengthened it at the head?'

'I didn't. You misunderstood. Oh, shut up.'

Then I retreated to the foredeck, where I sat crying.

The East Australian Current (EAC) flows in a southerly direction offshore, and a counter-current flows north closer to the shoreline. Sailors use these currents to their advantage whenever possible. But they are not like two rivers flowing in opposite directions. Rather, they flow in very large eddies, even in circles or semicircles, which are not fixed in location or strength. As climate change progresses, we have noticed alterations to the pattern of currents in the Tasman Sea. They appear stronger and the patterns seem to reach further afield.

The wind soon returned. We were into the temperate zone and a southerly front followed the westerly winds. The wind changed from five to thirty knots in thirty seconds and, in preparation, we were already reefing for the front's arrival. We were 165nm from Coffs Harbour and hoping we could reach port before the next front which was forecast to arrive in two days. The current was against us by a half to one knot as we came in closer to the coast, which meant slow progress when the wind was in the easterly quarter. Progress improved as the wind came round to the north-west and we arrived in Coffs just over seven days after leaving Nouméa. Coffs Harbour was the most convenient place to clear Customs back into Australia. We also needed to buy fresh provisions and shopping is easy there.

There was news from Jamie. He wanted to fly home for Christmas before sailing *Possibilities* back down the east coast of South America in 2007. Despite an argument with Laura, he hadn't given up his plan of sailing to Tierra del Fuego. We were also in contact with David and eager to meet his new love. It appeared we'd have a family reunion for Christmas.

After two busy days in Coffs Harbour, we set off for home, two knots of the south-going current helping us on our way. At times we were covering the ground at just under nine knots. We stopped for one night at Camden Haven before continuing, this time sailing overnight. It was another brilliantly starry night with a full moon.

At dawn the wind faded and we motored over the Swansea Bar at 5.30am and through the bridge at 6.00. We were satisfied with both ourselves and with *Osprey*. She'd performed well and we'd recovered from our fear she'd roll over again.

⚓

Heather Stevens breezed into our lives at the weekend (she was working in Sydney) and we were delighted to welcome her. She was bright and cheerful, a ray of sunshine. We were happy for David. He had been busy working on his concrete ammunition barge, turning it into a home and now he'd fallen in love with a girl to share it with him.

Jamie arrived home, too, staying for much of the summer before returning to *Possibilities*. It was lovely to have both our boys about and especially good to have them both so happy. Jamie was happy hanging out with Laura Anthony from the Hunter Valley.

Within a week of arriving back at Lake Macquarie, I was out house-hunting. We had decided we would sell the unit in Hornsby and find a place to live in Rathmines. It took until mid-winter the following year (July 2007) to settle on a house – a small fibro cottage close to the most sheltered corner of Lake Macquarie, where we could moor *Osprey*. The size of the house and the location were perfect, so we accepted the fibro-asbestos materials as a necessary compromise as well as the extra-large sized block of land. In the meantime, we continued living on the boat.

The necessary renovations took several months. We gutted the old kitchen, opened up a decent-sized window, installed a second-hand kitchen, and repainted the whole of the interior. The last big job was to sand and polish the hardwood floor boards that had been hiding under a very old and scungy carpet.

In March 2007, David and Heather became engaged to marry. Heather immediately began organising the wedding. As a result of all this planning, David insisted on bringing the wedding date forward to October 2007. He thought six months was more than enough time for planning, but they had to leave the honeymoon holiday until late winter 2008 when they could afford the fares to China and Thailand. By that time Heather was pregnant; and their son, Tasman, was born in November 2008. I was overjoyed with the birth of my first grandchild. Ian was very happy too.

So much family activity meant that we didn't want to be far from the action. For our next trip the following year (2008) we decided to stay in New South Wales. We thought that a longish river voyage might be an interesting thing to try next, and set about preparing to do just that.

10 Up the Clarence River

Many CCCA club members had, in the past, enthused about a voyage up the Clarence River in northern NSW. We thought that would be an ideal trip for 2008 – not too far from our home at Lake Macquarie, but a completely different experience to anything else we'd done.

We checked and found we already had most of the charts for the northern NSW coast and the river itself was detailed in Alan Lucas's book, *Cruising the NSW Coast*.

In preparation, Ian resealed all the windows on *Osprey* that had developed minor leaks and we completed a dozen other minor maintenance tasks, ensuring that all systems were functioning optimally and that spares were stowed. We moved on board three nights before our departure in late May in an attempt to minimise my seasickness once we were out at sea.

The Swansea Channel between Lake Macquarie and the ocean is tidal even though there is no tide in the lake. One normally leaves on a rising tide, stopping at the moorings either side of the Swansea Bridge. Slack water is the safest time to cross the bar, otherwise tidal outflow and onshore waves and swell can combine to cause dangerous breakers.

At that time, dredging of the channel was usually to two metres. Sand drift into the channel is a perennial problem and dredging is normally carried out prior to Christmas *if* a dredge is available. It was due to take place again in a few months. In the channel that day, *Osprey* went hard aground and we needed Waterways to tow us off. For a ten metre yacht, *Osprey* has a fairly deep draft – slightly over two metres when laden, as we were, with full tanks of water and diesel, and enough food on board for our cruise. We were finally at sea by midday on the 27th May, with sails set and heading north.

We were in an adventurous mood, keen to explore anchorages along the coast and not sail overnight. Our first anchorage was in Fullerton Cove off the Hunter River at Newcastle, where we were well north of the commercial port traffic. We hadn't been into the port by boat before and I was surprised to find the area around the Cove undeveloped and very quiet.

The next day, we continued to Port Stephens. We spent the first night on a public mooring just outside the Nelson Bay marina and several hours in the marina the next day (my birthday) with the batteries hooked up to the charger. It was pleasant not to suffer from seasickness on my birthday, not that we did anything special that day except go for a stroll along the shady foreshore.

After an overnight stop at Sugarloaf, a fairly open anchorage on the coast, we motored on in very light conditions to Cape Hawke, into Cape Hawke Harbour and up to Forster. We'd been there before and knew about some of the dangers of that waterway, especially at peak tidal flow. Our visit coincided with one of two annual king tides. We had anticipated only two days there; we stayed several – and experienced a nightmare.

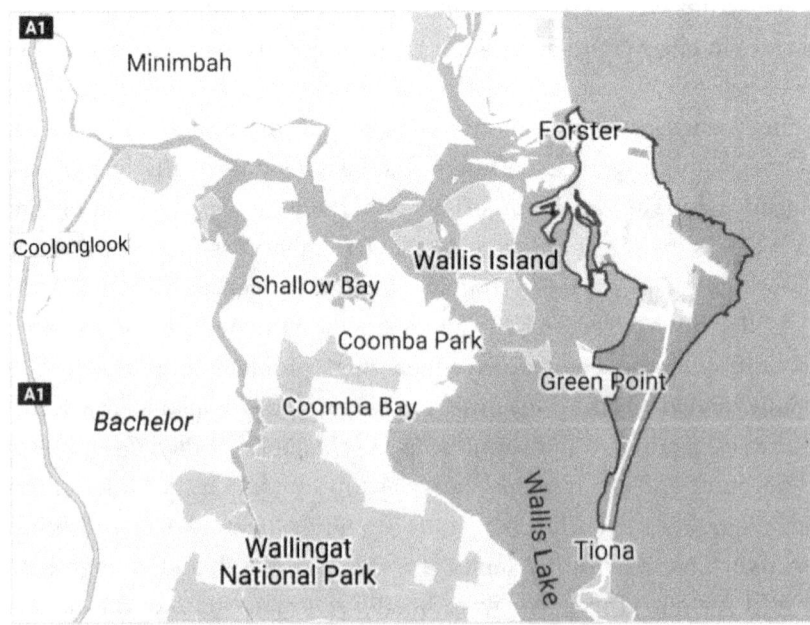

Fig 4. *Map of Forster & Wallis Lake, NSW*

Wallis Lake, fed by two rivers, lies upstream of Forster. This lake provides a large body of water for the outgoing tide and after a couple of very calm days, the wind turned northerly (abeam of *Osprey*) and began to increase, while, at the same time, the tide was pulling us downstream. There is a sandbank beside the channel which is over-topped only at a king high tide, which was due. Before bed, Ian had placed the second anchor and chain on deck ready to deploy in the event of *Osprey* dragging.

Just before midnight, we awoke to a changed motion. When I looked out and saw what was happening, my adrenalin shot skywards and my heart began pounding. The opposing wind and the force of the tide had swung the boat sufficiently for the anchor chain to catch around the keel, holding us sideways, the wind pushing us towards moored boats. Then the anchor began to drag. We hastily donned wet weather gear and started the engine. Ian's description, written for our club magazine, tells it all:

A Bit of a Drag at Forster in *Osprey A*

We were anchored with a 35lb *Trefco* plough and 30m of 5/16 inch chain mid-channel between the Fish Co-op and the bridge. On June 2nd at 2315hrs, with the wind against current, the anchor chain passed behind the fin keel. With the yacht held broadside to the current of about four knots, the anchor dragged. Attempts to unwind the anchor chain under engine were useless. We dragged at about 1-2 knots down the channel. The yacht, still held broadside and port side to the outgoing current, surged fore and aft across the channel about 25m, threatening to crash into fishing boats, yachts and the rock wall.

The only control we had was to put the engine hard astern when a collision was imminent. We had a second anchor ready to deploy, but the cross channel movement of the yacht was so fast and erratic that we were not sure we could place the anchor so that we would be clear of the moored boats. After dragging 600m past all the pile moorings, an eddy held us towards the sandbank side of the channel for about 30 seconds and I dropped the second anchor, a 27lb *Manson* plough with 18m chain and nylon warp. It held. We retrieved the main anchor without difficulty.

While dragging down the channel, Jan radioed the Volunteer Marine Rescue (VMR) and we were told that a vessel would come to assist. After the second anchor held we considered cancelling the request for assistance. However, since the process was in train and there was still a small chance that something else might go wrong which might prompt another call for assistance, we did nothing.

The VMR vessel arrived, passed a line, and carefully towed us upstream, the second anchor dragging behind. At a wider part of the channel, they held us steady while I retrieved the second anchor. They then towed us to the Fish Co-op and held steady while we moored to the piles. We are very appreciative of the care, skill and patience of the skipper and crew of the VMR vessel.

Remarks

It was good that we had a second anchor ready to go. Not so good was that it took us so long to realize we were dragging, by which time we had already dragged about 100m and were close to the Fish Co-op. We had heard the anchor chain rasping on the hull, but I did not go out to check, confident that the GPS alarm would warn us if we dragged. The alarm did not go off. I can only assume that I had forgotten to set it. Jan, made uneasy by the rasping chain, went on deck to check what was happening.

The streetlights cast a deep shadow behind the rock wall, so I could not see well from the cockpit. Jan spent most of the drag at the bow, yelling back to me to go astern when a collision was close. I yelled back for her to keep her body parts out from between. We did hit one yacht, but so lightly that there was no damage.

The following night we remained alongside the Fish Co-operative piles. At high tide, water covered the sandbank to windward and the motion was quite uncomfortable. At 0200 the fender board snapped. Luckily I was standing beside it and jammed our last and largest fender down between boat and pile.

⚓

We were still feeling tired and demoralised the next morning. The next high tide was predicted to be as high as the previous night's and the barometer was plummeting, indicating worse weather to come. I wasn't sure what we should do, but Ian had a plan. It was no use anchoring again because, at high tide it was highly likely the anchor chain would wind around the keel again.

At slack water (low tide), Ian untied our mooring lines and we motored downstream to some piles, where we tied up. This wasn't entirely satisfactory either, and when the tide reached its highest in the middle of the night there was a tremendous strain on those lines. Ian added a second pair of lines to the poles. We stayed there until slack water the next morning when we thought it was safe to depart for Coffs

Harbour. Once we were over the bar, our tension dropped away, we relaxed and enjoyed sailing north.

Out at sea, the wind was south-westerly off the land and, consequently, the sea was calm. Despite current against us, we surged along. Once around Crowdy Head, the current turned with us, so we continued making good over the ground despite the falling wind.

In the early evening, the lights of Port Macquarie showed abeam, and much later, we passed Smokey Cape. Before daylight, we were off Coffs Harbour and Ian made the decision not to stop, but to keep sailing north. When the wind dropped to less than two knots in the early hours of the morning, I regretted not stopping because I like Coffs Harbour.

After breakfast we saw three humpback whales pass us, just quietly swimming north, and we continued in their wake. We motored, motor-sailed, and were able to turn off the engine briefly and just sail until we entered the mouth of the Clarence River in the late afternoon.

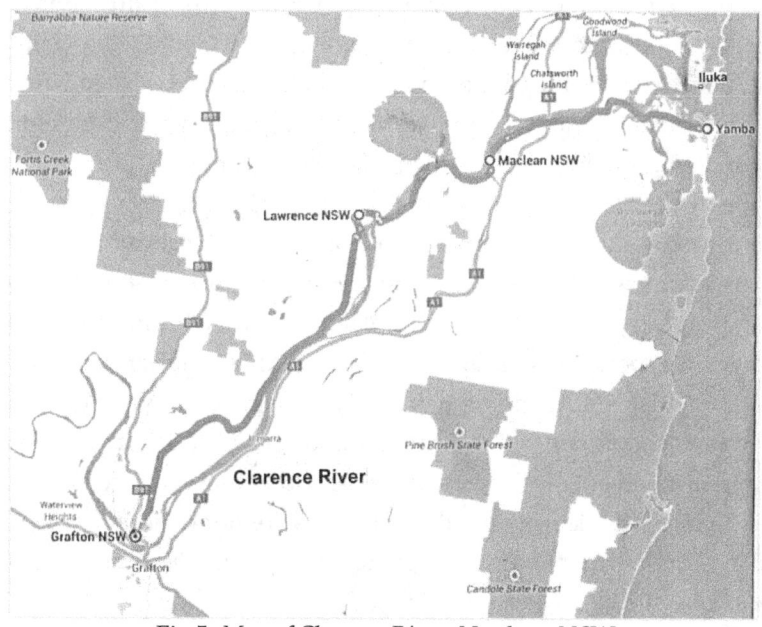

Fig 5. *Map of Clarence River, Northern NSW*

It was dark by the time we made our way into the Iluka anchorage on the northern side and found a place to drop anchor. Initially, I felt extremely relieved that we had arrived inside the river mouth without any physical damage to us or our boat, but there is something addictive about the adrenalin rush that comes with the adventures we have experienced. There is also a satisfaction in extricating ourselves safely from hair-raising situations like the one at Forster, or the dismasting in Bass Strait, and even in getting over seasickness, so that I can relax and enjoy the sailing.

We had been to Yamba on the southern shore twice before, but we'd never visited Iluka, so we spent the weekend there, walking about the village and into the coastal nature reserve on North Spit.

On Monday, we took *Osprey* across the channels to Yamba Boat Harbour, hiring a berth at the marina. There, we were able to catch up on emails, top up fresh food, water and fuel, and do our laundry. Yamba marina is the only one I know on the entire East coast which has an outdoor clothes drying area. Sun-dried clothes are healthier.

Back out on the river, we booked the Harwood Bridge, which carries the Pacific Highway over the Clarence River, to open for us. The original bridge, opened in 1966, is a truss bridge with a vertical lift navigation span. It is identical to another truss bridge opened in 1964, the Nelligen Bridge, which crosses the Clyde River near Batemans Bay, also in NSW.

Construction began on a replacement bridge at Harwood in January 2018, and the new bridge is due to be completed in late 2019. Coincidently, the Nelligen Bridge is also being replaced.

A few miles west of the Harwood Bridge lies Maclean, the 'Tartan Town', which has a town-wharf with free berths, water and electricity, making people in boats feel welcome. All through the business district of the town, the power poles are decorated with the tartan patterns of different Scottish clans, giving a friendly atmosphere. Typical of small rural towns, people are willing to stop and chat. I really like this little town and enjoyed our walks there. Houses and gardens are well kept and the whole town district is very tidy. The locals are proud of their town.

The Clarence River runs about sixty-five kilometres from Yamba to Grafton and further. It is a wide and full river, slowly meandering through farmland in northern NSW. The tidal influence continues a considerable distance inland, and we needed to take that into account while motoring upriver.

This river cruise was quite unlike anything we'd done before and I felt almost dreamlike during the few days we spent there. At one anchorage, I remember, we awoke to damp freshness and mist. As it cleared, a few birds came into view on the water and then I saw an overgrown island. There was no noise and nothing else in sight, yet there are farms along the riverbanks, with cows, dogs and farm machinery. The road to Grafton runs along the south-eastern bank, but that morning, all was still.

A couple of days later, we reached the small city of Grafton, where the highway bridge prevented us from going further upstream. On the river, there was rarely even a hint of traffic fumes. We stayed for a couple of days and then felt inclined to move on. I have no memory of why we didn't walk far from the riverbank and didn't go into the city. Maybe we just wanted to isolate ourselves from hustle and bustle.

Going downstream took us only two days. We stopped overnight again at Maclean, and returned to the marina at Yamba to refill our diesel tanks before heading back to Iluka, staying there for several days. During our stay in Iluka, we finally made the acquaintance of Alan Lucas and his wife Patricia. Alan is, of course, the author of the cruising guide we were using and famous in Australian yachting circles. We enjoyed chatting to Alan and Patricia, talking about mutual acquaintances and places we'd sailed. Patricia took a couple of photos of *Osprey* and sent them to us later. We were thrilled by that.

Osprey *departing the Clarence River (Photo by Patricia Lucas)*

We set off for home on Monday morning, 23rd June. During the three-and-a-half day voyage back to Lake Macquarie the wind was frustratingly changeable just as it had been on the way north.

Sometimes we were able to sail and, at other times, the wind dropped away and we used the motor. Ian frequently pulled up and down the sails. Reefing and poling out sails are tiring activities when performed all on one day. We weren't able to set the rig and relax. This kind of thing frequently happens during coastal sailing, which is why we prefer to cross oceans.

By contrast, as we were crossing the Newcastle Bight, notorious for bad weather because it is relatively shallow, there was a strong blow, and then we had to try to hold the boat back, so that we didn't approach the Swansea Bar too early. Soon after dawn, when the tide was slack, we were able to cross the bar and motor up to the Swansea Bridge. Later in the morning, the Coast Guard helped us through the shallow patches of the channel to the drop-over into Lake Macquarie.

It had been a very short cruise that year. We'd been away for barely one month, but there was David's wedding being planned for October.

After picking up our mooring and unpacking the boat, there was the usual contradictory set of feelings: satisfaction in a voyage accomplished and the bringing of ourselves and the boat safely home, against the regret that there would be no more voyaging at sea that year. Ian always felt that more keenly than I did. He was always happiest sailing at sea.

Soon after David and Heather married we began planning our next voyage. Would we have another international adventure?

11 New Zealand's East Coast to Stewart Island

Latitude 41° South. Hobart? No, although it was nearly as cold. We were in Nelson, New Zealand. We drifted across Tasman Bay during a sleep-deprived night, and at dawn we were well rugged up to motor into Nelson, the closest port of entry. It was nearly thirteen days since we'd left Newcastle, NSW, where we farewelled family and cleared customs fifteen days before Christmas, 2009.

When we left Newcastle on the ninth of December, the weather was warm and almost calm. We'd cleared the harbour under full sail and slowly wended our way through the ships at anchor outside the port. Those were the last ships we were to see before we rounded the Farewell Spit into Tasman Bay at the north-western tip of the South Island of New Zealand.

When I hadn't been seasick by late afternoon, my hopes rose that I might be spared the misery. It was not to be. With the coming of evening, the wind began to rise and so did the swell and seas. By nightfall, the dreaded *mal de mer* was upon me, not to relinquish its grip for four more days.

I rose from my bunk to go to the toilet before settling down for the night. The weather was hot and humid which made my hands sweaty. Part way through the act of standing up, I misjudged the boat's motion and, as my handhold slipped, I was flung backwards across the bunk striking the back of my head on the new teak handhold under the window. Ian had been standing close beside me at the time, ready to steady me to the toilet, but my fall came without enough warning for him to help prevent it.

I felt angry that I had sustained another head injury less than nine months after falling and striking my forehead onto a concrete floor during a yoga class. My full body weight was behind that fall which affected my ability to control my impulsiveness.

Counting the accident in Bass Strait, this was my third concussion and consequent brain injury. My head felt woozy and I was unsteady for the next three days. The wind rose for two days until it was over thirty-five knots in gusts from the south-west. Our course was south-easterly, so this put the apparent wind forward of the starboard beam. Although the seas rose with the wind we were not unduly uncomfortable, and I was very careful how I moved about the boat while *Osprey* made good way over the ground at between six and seven knots with a little help from some current.

Early on Saturday morning we were becalmed. The very light conditions lasted for about twelve hours yet by nightfall we were under full sail again, the big genoa poled out to starboard. Just after midnight, Ian went on deck and dropped the mainsail to keep our speed at a comfortable six knots, as too much boat speed with the wind on the beam does not allow the windvane to steer a proper course. A reduced sail area allowed the *Aries* windvane to do the hard work of steering. 'Harry' is the perfect crew member who never complains; without him, Ian and I couldn't sail with just 'two-up'.

Our progress was good for several days and mostly in the right direction. I had expected fairly warm conditions after the very hot six weeks we'd experienced before our departure from Lake Macquarie. Instead, the temperature didn't rise above the mid-twenties and became colder by the day. We pulled out the bag of winter clothes I had stowed in the port quarter berth and put on thermal underwear at night. I thought of the range of shorts and light summer shirts we'd brought, and wondered if we had enough warm clothes with us. On Monday, a small front passed, bringing more cold wind from the south.

The wind eased slowly and Tuesday, Wednesday and Thursday saw us mostly becalmed again. By then, I was over the seasickness and concussion. Ian pulled the multi-purpose spinnaker (MPS) out from under the vee berth. He raised it, trying to make some headway in the light conditions and when the wind dropped altogether, we motored for a few hours.

I couldn't help but compare this crossing of the Tasman with the trip we'd made in December 1989, in *Realitas*, our Phantom 32. On that trip too, our aim was to have Christmas with members of my family in the South Island. Although *Osprey* sails, on average, faster than the Phantom, on that occasion we had only one gale and about eight hours of calm on a ten-and-a-half day voyage. During that calm it was hot enough for us to swim over the side. This time, unless we were keeping watch or there was work to do outside, we stayed in the cabin to keep warm. Wind patterns are changing with climate change, and over the years since this voyage, I have noticed more and more instability – easily checked in these times of real-time viewing on the internet.

I had no temptation to tow a fishing line. Apart from an albatross and a couple of shearwaters each day, there was no other life to be seen. Although the birds circled and searched, we never saw them catch a fish. The ocean seemed as empty as I've ever seen it; just a wide expanse of water with our small 10-metre yacht sailing upon it. I thought about the gigantic fishing ships with huge nets that scoop the fish from the ocean depths and process it at sea. They are denuding our oceans of sea life.

Late on Thursday, the barometer began to fall slowly from 1020hPa down to 1008hPa by Sunday morning. Over the next two to three days, the wind increased to about thirty knots during one or two small squalls. By then, we were into the weather forecast area from New Zealand and their very detailed high-seas forecasts. They predicted a front of thirty-five knots, but instead of the wind easing after the front had passed, the wind continued to increase until we were experiencing gale-force winds and rising seas.

On Sunday, I began to feel seasick again. By the time I was passing buckets to Ian for cleaning, he realised it would be easier to stop sailing. He was getting tired as he'd been doing all the sail handling and navigation himself. Most modern yachts won't heave to, and *Osprey* doesn't do it very well either even though she has a classic wineglass-shaped hull but the keel, though deep, is fairly short.

Ian pulled the triple-reefed main in hard and furled up the genoa. Under these conditions, *Osprey* made about one to two knots of forward motion. Still being sufficiently clear of the New Zealand coast and with no shipping in sight, we were able to rest easy while the forty-knot gale blew itself out after about twelve hours.

When daylight broke on Monday, 21st December, conditions had eased sufficiently to get underway again, the wind having backed to the south-east. Ian tacked the boat and, close-hauled, we set course for Cape Farewell. First, we sighted the high hills to the west of Nelson. Then the sandy spit of Cape Farewell gradually emerged over the swells and, by shallow Golden Bay on the western side of the much bigger Tasman Bay. As usual, when I approach New Zealand, whether by sea or air, I experience an excited tightening of my chest muscles in anticipation of setting foot in my homeland. Even though I have lived well over half my life in Australia, it doesn't feel the same as being back where my earliest and most formative memories exist.

We called Nelson Harbour Control and informed them of our approach, arranging to meet Customs shortly after daybreak. From Farewell Spit across Tasman Bay to Nelson is roughly sixty nautical miles. The forecast was for light and variable winds in the bay. Under full sail, we slowly drifted towards Nelson. No longer seasick, I took as much of the night watch as I could manage, it being cold enough at eleven degrees to keep me wide awake. At 3.00am, I woke Ian. He started the engine and we motored for the harbour entrance, still about ten miles away.

With directions from Harbour Control, we found our way into the customs berth, where the Customs Officer, dressed in shorts and a short-sleeved shirt, was awaiting our arrival. I shivered and felt slightly ridiculous in thermals, jeans and skivvy, jumper, full waterproofs, fleece hat and neck warmer – but dammit, I was cold.

The customs officer seemed very suspicious. 'What took you so long?'

'We thought we were being polite, arriving in daylight,' I told him.

'Have you been into Nelson before?'

'Not since 1989, in a different boat,' I replied.

After the customs officer left, I congratulated Ian on getting us to New Zealand in time for Christmas with my 'baby' brother in Blenheim, just a short bus ride away. It was time to cook up a big breakfast and eat it before the Quarantine Officer arrived at eight o'clock to take away any fresh food we had left.

My brother, Greg, picked us up later that day and drove us to Blenheim where he was living at the time. We stayed with him for several days over Christmas. I am fond of my younger brother and it was good to catch up with him. We had not been together since he had split up with his wife. As we drove, I noticed that the countryside had changed since my last visit. The Wairau Valley, once mixed farming country with sheep, dairy and crops, was now dominated by vineyards and wineries.

Ian wanted to bring *Osprey* east to Picton and Greg was keen to sail with him, so they decided that Greg would sail from Nelson with Ian and I would look after Greg's car and Zoe, his dog.

They experienced almost no wind during the first afternoon and, after negotiating French Pass, they anchored in Deep Bay. The next day, they sailed through Cook Strait and up Queen Charlotte Sound to Picton.

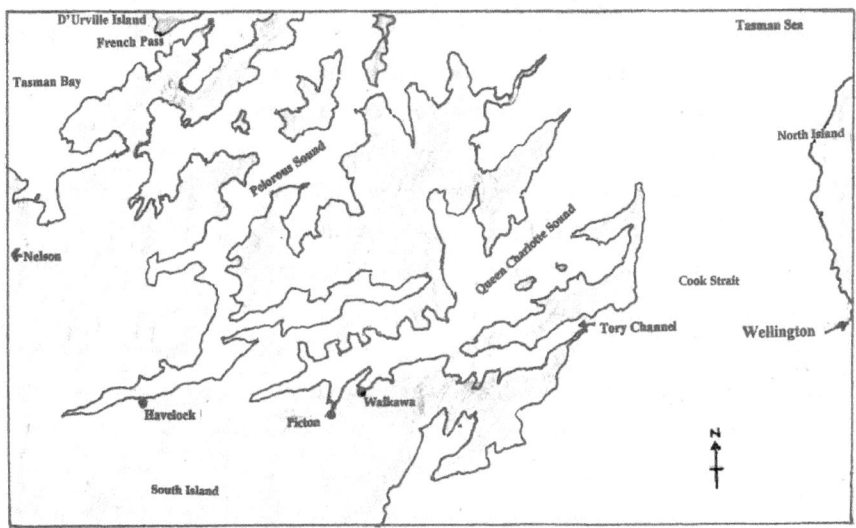

Fig 6. Map of Marlborough Sounds

My youngest brother, Greg Hormann.
Greg was killed in a vehicle crash at Tekapo in December 2017

Ian had not expected the marina to be full, but it was peak holiday season and he had to leave *Osprey* on a mooring at first, there being no berths available. Later that afternoon an empty berth became available and I arrived just in time to help the men secure the boat. We stayed on board and Greg went home with an invitation to return for dinner with us the next night, New Year's Eve, and to watch the fireworks afterwards. As to be expected in a small town, the display was nowhere near the standard of Sydney where millions of dollars are spent to entertain the international crowd, but it was a pleasant evening nevertheless.

The day after New Year's, we met a very friendly couple, Ros and Pip, on *Isis*, a steel forty foot yacht from Whakatane in the North Island. They had cruised extensively on the east coast of New Zealand and were keen to advise us about the best places to visit. They had recently sailed *Isis* to Stewart Island, and were full of enthusiasm for the place. Pip eagerly offered to lend us their cruising guide for the island which lies in the Southern Ocean, across Foveaux Strait from the South Island. Pip and Ros's attitude was infectious and soon we were eagerly planning to follow where their keel had led.

We really enjoyed meeting this couple, quickly becoming very friendly as you do when cruising. While we were having a cuppa with them, we watched some *Optimist* dinghy-sailing going on close to the marina. Wind gusts were up to thirty knots, and it was plain to see why New Zealand sailors are so adept in heavy weather when kids start out like this. The wind gusts later increased in strength and, in a particularly strong gust as our boat lurched against the pier, one of our large fenders burst. This was a hint of things to come.

Greg suggested we check out Ngakuta Bay where we could borrow a mooring from a friend of his. My brother assured us the bay was sheltered. We set off for Ngakuta late morning and picked up a vacant mooring.

Ian took the dinghy ashore to find Greg's friend and identify which was his mooring. By the time we were securely tied on, the wind was increasing. Ian decided that the outboard and dinghy should be on deck. It was just as well he did. The wind came in from the north-west at about thirty knots for the remainder of the afternoon. During the night, it buffeted *Osprey* with every gust so that she swung wildly about the mooring.

The next day's wind was even stronger. We stayed on board watching sheets of water lift off the surface with wind gusts of over fifty knots (92kph!). We were reluctant to put our heads outside because of the flying water and salt spray. Even the locals weren't keen to swim in those conditions although they had been jumping off the pier the day before.

The gale eased considerably by the following morning and cabin fever was setting in so, after breakfast, we decided to go ashore for a walk. We thoroughly enjoyed ourselves stretching our legs along the road out of the village. The gravel road wound south-westerly and the hill alongside protected us from the wind. On the seaward side, there were native shrubs and grasses. We were gone for over an hour and when we returned to the dinghy we found the wind blowing a steady thirty knots, gusting to thirty-five.

Ian tried to row us both directly into the wind but it took us backwards faster than Ian could row forward and he tired very quickly. He managed to get us into the lee of the public pier and tie on so that he could rest. Ian thought that on his own he would be able to row out and collect the outboard motor then come back for me. I waited on the pier and watched anxiously. Ian managed to catch hold of a mooring where he rested for a short while before striking out again for *Osprey*. The wind was still increasing and the dinghy hobby-horsed wildly while attached to the leeward side of *Osprey*. I feared that Ian would fall overboard as he stood and brought the outboard off its mount. I bit my knuckles as he secured the outboard to the dinghy transom, started the motor and released himself from the yacht. It was only seconds before he arrived downwind at the pier.

Gingerly, I climbed into the heaving dinghy. We took a lot of spray over the bow as we motored into the wind, so that we were both soaked by the time we arrived on board and my lips were salt-encrusted.

Ian handed the outboard up to me and I secured it on its mount on the cockpit railing; to leave it on the dinghy behind the yacht was asking for it to flip and drown the little motor. If that was difficult, it was harder still to bring the dinghy on board with the wind trying to send it shoreward. Finally, the dinghy was firmly secured to the deck and we could go inside, strip off, wash and put on clean, dry clothes.

That had been the westerly, prior to a cool change. When the southerly finally arrived, the land provided us with shelter and we experienced a calm, quiet night.

The following morning was still calm and we returned to Picton, tying up on the public wharf while I shopped for groceries and ordered vacuum-packed meat from the butcher. There was a rule against our staying on the visitors' wharf overnight, so we motored around to Lochmara Bay, where rain and thunderstorms kept us inside for the remainder of the day. However, it was fine the next morning, so we took the opportunity to move to Onahau Bay where we went ashore for some exercise.

The weather that January followed a pattern of blowing a gale from the north followed by a gale from the south over each 24-36 hours. Even the locals were complaining. We knew that if we sailed south to Stewart Island, it was highly likely we would, at some stage, have a gale on the nose. We decided to do two things while waiting for the weather pattern to change: One: Hire a car for a week to visit family members who lived in the South Island. Two: Do some more alterations to *Osprey*.

Ian had become impatient with dropping the anchor two or three times before it held. He told me he had decided to buy a *Rocna* anchor, which he thought would provide more secure holding further south than the plough-style anchor we had been using. However, a *Rocna* wouldn't fit on our current bow roller and we needed to alter the fitting.

At Picton, we made arrangements to hire a car and then motored *Osprey* to Waikawa, the nearby industrial suburb where we took a berth in the marina. We ordered the anchor and accepted a quote for a new bow roller. Ian had designed what he needed and the fabricator promised to weld it up and fit it onto the foredeck while we were away.

There was already a pad-eye fitting on the foredeck for an inner forestay. Ian had never used that fitting because he deemed it too weak. He wanted to strengthen it. This would allow us to rig *Osprey* as either a cutter or to use the storm jib on the inner stay. Ian eventually figured out a way to reinforce this fitting himself.

Fig 7. *Map of NZ including Stewart Island (inset)*

We set off by car to visit my two older brothers, Alister in Ashburton and Peter in Wanaka. Two of my nieces, a nephew and Alister's grandchildren also lived in Ashburton and we visited them all. I always enjoy catching up with my family.

On Sunday evening, we stopped off with an old school friend of mine. There was a free concert that evening at the Sound Shell on the beach at Timaru where I had lived as a teen. We went to the concert and, as we listened to many songs from the 1960s, Brenda and I reminisced about old times and the mischief we got up to together during the summer before we left home for university. What a night for memories!

When we continued south towards Wanaka, I revelled in the greens of the countryside of my youth, so different from Australia. We took the 'scenic route' that was a series of back roads when I was a kid but has been upgraded into a major tourist highway.

In Wanaka, we stayed for two nights with my eldest brother and his wife because Peter wanted to take us out on Lake Wanaka in his speedboat. That was a very different kind of boating for us. We zoomed north for half an hour, spray flying out behind, and came to a halt at a small island nature reserve, noted in particular for both wekas and wetas.

The weka, like the kiwi, is a flightless native bird; a weta is like a very large grasshopper or cricket-like insect from the age of the dinosaurs and is endemic to New Zealand. The giant weta is the largest insect on earth today. They are endangered because of rats and are near extinction. We didn't spot a weta on our walk to the top of the island, but on the way down, a shy weka deigned to peep through the undergrowth at us.

Peter died, aged 77, in 2018 not long after he and Barbara moved to Geraldine in South Canterbury. I prefer to remember our visit to him in Wanaka before Parkinson's and dementia took hold.

On our return journey to Picton, we surveyed the port at Timaru in case we needed to stop there with *Osprey*, and then broke our trip again with Alister and his wife, Bev. Alister died of liver cancer in 2013 and Bev died a couple of years later. I am so glad we spent time with them. Inside six years, all three of my brothers passed away.

After visiting Alister's family members, we drove on north, reaching Blenheim five and a half hours later. A very brief stopover at Kaikoura convinced us we didn't want to ever try entering that tiny fishing harbour. Outside the harbour the water was strewn with rocks. I tried to envision entering that limited space in the dark. No way!

On our return to Waikawa, Ian inspected the new bow roller. He pronounced the fitting a success and installed the recently purchased *Rocna* anchor on it. There was still the inner forestay attachment point to reinforce. When Ian had recovered from the car trip he installed reinforcing under the deck for the pad-eye bolts and connected the deck at that point to the anchor-locker bulkhead. We had to take out the deckhead lining to do the job and then replace it. That completed, we felt reassured that the pad-eye was strong enough to use for attaching an inner forestay that would mostly be used for the storm jib in strong winds. I enjoyed helping Ian on boat maintenance. I lay on my back on our bed, feet holding up the ply lining, while Ian screwed it back in place.

Waikawa Rigging made baby-sized stay strops for the storm jib, our charts arrived, but we couldn't leave because the new batteries on order and my replacement Visa card from Australia were yet to arrive.

We finally moved out of Waikawa on February 9th, motoring to the Bay of Many Coves where we had arranged to meet up with the parents of Lisa, Jamie's new partner. Jeanne and Al had been bushwalking and camping on the western side of Golden Bay and were now staying at a resort with some English friends. They came out to *Osprey* in kayaks and we spent a delightful couple of hours with them. Although not sailors, this couple were fit, adventurous and accepting of our lifestyle. We liked them a lot. By then, Jamie and Lisa had been together for two years and looked like becoming a more committed couple than either of Jamie's previous long-term girlfriends had.

A weak cold front came in overnight and the next morning with the wind predicted to move into the south-east, we decided to leave to sail south. Several hours later we had packed up the boat into ocean-going mode and raised the anchor. There was almost no breeze, so we motored all the way, crossing Queen Charlotte Sound into the Tory Channel which would take us via a shorter route to the east coast. This is the channel used by the inter-island ferries. (See *Fig 6*. Map of Marlborough Sounds. p.193.)

Out on the open ocean we unfurled the genoa fully to make the most of the breeze, and headed south. Apart from the mild sea breeze the following day from 11.00am to 6.00pm, we motored, sighting some albatross and sooty petrels. For a change, Ian was nauseous and I wasn't.

Only a day and a half after leaving the resort where we'd met up with Lisa's parents we arrived off Godleys Head on Banks Peninsula at the waypoint Ian had set. We planned to enter Lyttleton Harbour, but first we anchored just around the corner from Diamond Head in Purau Bay. We had stayed there in a friend's holiday house back in 1997, after my mother's death.

Ian called Harbour Control on Lyttleton Radio. They arranged for us to tie up on a floating pontoon at Dampier Bay Marina at the western end of the commercial harbour.

We were made very welcome and offered free mooring overnight as well as access to showers, toilets and laundry.

My friend since student days in Christchurch, Marjorie who lives not far away in Mount Sumner, came down to visit. She was very obliging, running us about in her car to buy more groceries, meat and diesel because we'd already used much of the supplies I bought in Picton. We shared a meal with her and her husband John whose company we really do enjoy. Like Ian, John is an engineer, but in electrical engineering at the University of Canterbury. It was through a Sydney connection of John's that I had first met Ian.

A steel yacht called *Spellbound* had been moored behind us when we arrived in Lyttleton. They departed for Nelson the day before. Now, here they were returning and one of the crew called to me from the deck as I stood on the dock.

'Our engine won't start.'

Spellbound was approaching under sail; they had left it too late to turn. Their sails were limp. Chad heaved a bow line to me. I grabbed it. Flipping it around a cleat, I pulled the line in rapidly. The keel stopped in the mud. The bow was just short of the rocks. The skipper jumped into a tied-up dinghy and from there onto the dock. Together, we hauled the boat out of the mud, turning her stern to stern with *Osprey*. As *Spellbound's* crew secured her to the dock, my adrenalin level and heartbeat begin to subside. Their entry into the marina could easily have become a fiasco involving rocks on one side and *Osprey* on the other.

The crew of *Spellbound* bundled the mainsail untidily on the boom. Someone brought up the mainsail cover and struggled to tie it over the lumpy sail.

As I retired inside *Osprey*, I overheard a comment.

'This is like putting a corset on a fat girl.'

I subsided in laughter, looking forward to relating the incident to Ian when he returned.

We pulled out of the marina next morning, satisfied we had enough stores on board, and returned to anchor overnight in Purau Bay before motoring out to sea early the next morning for the very short trip around to Akaroa. I had wanted to sail into Akaroa Harbour since I first looked down on it from the hills when I was a university student in Christchurch. Now, we were there in our own yacht.

In 1840, the British learned that a ship of French colonists was due to arrive, but only days before their landing, the British rushed to declare sovereignty, raising the British flag. The French ship landed and the colonists became British citizens. French street names and some French surnames are the historical evidence of the French settlement here.

There was no space at the town wharf to tie up, so we anchored *Osprey*, pulled the inflatable dinghy out of the forward cabin, and blew it up to row ashore. There we found a thriving tourist town, the wharf dominated by charter boats and paua shell jewellery stores. Paua is a kind of abalone. It was a pre-European tradition for the Maori to use the decorative shell to augment their carvings, especially for eyes. The making of paua jewellery is now a national industry. There was even a pearl farm for beautiful blue paua pearls in Akaroa Harbour, but when I enquired about visiting it I was told it had closed to the public not long before.

The charterers and other boating people were very friendly and welcomed us to use their yacht club facilities for showering and laundry. Even more welcome was a shelf of books to swap with ones we'd read. Ashore, there were more souvenir shops and plenty of cafés and restaurants. There was also a shop selling woollen clothing and we both bought merino tights and tops for the cold weather further south.

Three days later, we were sailing down the coast again, with enough wind to make me mildly seasick for a few hours. Later that night, while I was on watch, I spotted a cruise ship coming up astern. We were moving quite slowly and, as Ian had not long gone to sleep, I didn't want to disturb him. In the dark, my sense of direction often deserts me, but I thought I could get the lights right. I could see the ship's red light and reasoned that meant it would pass us to port, not starboard.

As the ship came closer, I started to panic and feel confused. I decided to move out of its way by starting the engine and turning to starboard. They must have wondered what sort of idiot was at the helm, because I was, in fact, turning into the path of the ship. They stopped. The ship then passed slowly to starboard, as I should have realised it would, given its lights. Port to port is when the ship is coming from ahead, not astern. Was this more evidence of my brain being damaged by three concussions?

Soon after we had passed the Otago peninsula (Dunedin and Port Chalmers) where Ian reported our position to Chalmers Radio, we began seeing and hearing fur seals. We mistook them for dolphins at first, but then we noticed their behaviour was different – and dolphins don't have whiskery snouts. Individual seals dive deeply for prey. They don't frolic in groups around the bow of the boat, and during the day, they can frequently be seen resting on the surface of the water, sometimes languidly waving a flipper. Once we recognised their barks at night, we became much more aware of their presence.

Despite it being mid-summer, temperatures dropped as we moved south and we became very thankful for our extra merino clothing. At times, on watch during the early hours of the morning, I resorted to filling a hot water bottle to warm up. This was like winter sailing, but it was the different experience I had wanted, so I wasn't complaining.

As we sailed east of Foveaux Strait during the night, various currents caused the sea to become irregular. The wind was also fickle and *Osprey*'s movement became jerky at times. I was very seasick and had to leave my valiant husband to deal with both sail changing and navigation. He stayed on watch all night. At times we were making five to six knots; at others, we were travelling under engine. By morning, we were in the lee of Stewart Island and the motion became regular again and pleasant, although the temperature was decidedly chilly. The hills were cloaked in dark green bush, right to the shoreline, there was little mist in the valleys that clear blue morning, and a flock of pacific gulls floated on the water. The air was invigoratingly sharp and fresh, until a large catamaran carrying tourists passed us, smothering us in diesel fumes before it turned into Half Moon Bay.

We were able to pick up a continuous weather forecast for this part of the Southern Ocean. Ian recorded hearing of a wind-gust of seventy-two knots at South West Cape (Stewart Island). At that point, I declared there was no way I intended to circumnavigate this island. We were already at 47°S – several degrees or about 445km further south than Hobart.

A couple of hours after dawn, we turned into Paterson Inlet. This is no small inlet, but in fact an extensive harbour containing small islands and many bays. The main settlement on Stewart Island is Oban, which lies just over a small rise to the north of Paterson Inlet at the head of Half Moon Bay. We anchored in Golden Bay on the southern side of that rise and, after breakfast, Ian caught up on sleep.

Seaward view of entrance, Paterson Inlet

Later that day, we rowed to the jetty and walked over the hill to acquaint ourselves with the services available in Oban. There seemed to be one of everything, but no mobile phone reception except for NZ Telecom. Internet was available at the pub, and I sent out a newsletter telling friends and family that we'd reached the limit of civilisation. There are no settlements south of Oban.

In such remote locations, people are friendly and keen to chat. Locals are proud of their town and happy to tell us about places of interest. One man told us about the bird sanctuary on Ulver Island in Paterson Inlet where he took groups of tourists. Having already seen a pair of tuis and some wood pigeons during our short walk, we decided to visit the sanctuary.

We awoke to brilliant clear skies the next morning and decided this was the right day to visit Ulver. The island was obviously the pride of the Department of Conservation employees. It was crisscrossed with well-made gravel paths and many trees had signs giving their botanical and common names. At the start of our walk we picked up a pamphlet showing the local birds and we managed to spot quite a few of them, including kakapo, saddlebacks, parakeets, oyster catchers and bellbirds. The robins were particularly friendly, but not as friendly as the flightless wekas on the beach. They had obviously become used to being fed by tourists and even climbed into our backpacks looking for titbits. We looked back to our trip on Lake Wanaka and laughed about the effort we went to just to see one weka.

A weka on the beach, Ulver Island

Osprey *Anchored off Ulver Island, Paterson Inlet*

In the middle of the day, the sky clouded over and rain began soon after. We hurried back to the boat. The weather forecast reported the approach of a strong south-westerly.

We moved much further up the harbour to find good shelter. On the chart Millars Beach anchorage looked good and, as we headed there, we were passed by a fishing boat that swung in towards the wharf. They beckoned to us to come over and tie up alongside them.

On board were Peter, the Kiwi boat owner, and Patrick from California. They were both professional divers and photographers, not fishermen, with sharks their main commercial interest.

The diving cage for watching white pointer sharks

Peter had designed the large cage on the back of his boat for tourists to view white pointer sharks at close range. I shuddered at the thought. No way would I go inside a cage and be lowered into shark-infested water, especially with a large gap at the top on one side.

'Isn't the gap here a bit big?' asked Ian.

Peter's response was to lower the side further.

'It's safe to have it down this far.'

The rain had stopped and Peter changed the subject.

'Have you seen the old whaling station yet? It's only a short walk from here. We'll show you the way if you like.'

The station was established by a group of Norwegian whalers in 1924 and abandoned a few years later. What had once been the mechanics of a thriving industry had deteriorated into scattered and rusting remnants of rails, barrels, try pots, winches, propellers, even the keel and a few ribs of an old ship were half-buried.

Back on board, the guys showed us a large cod they'd caught that afternoon.

'Would you like some for your dinner?' Patrick asked. 'I'm going to wrap it in foil and cook it on the barbecue.'

Shortly after we'd finished our dinner, the rain came in heavily and stayed all the next day. We weren't to see blue sky again during our time at Stewart Island, making us very thankful we'd visited Ulver Island in sunshine.

During periods of downtime when we were confined on board, I worked on my biography of Colin Kerby. I was still in the initial stages of the book, transcribing interviews I'd taped with him. (See *tinker tailor, soldier, sailor:* the life of Colin Kerby OAM).

When the rain eased, we moved back to Golden Bay where I picked up food we'd ordered, Ian bought more diesel and I did some laundry at the backpackers' hostel.

That afternoon, we sailed across to the southern shore of Paterson's Inlet and into Little Glory Bay, where there was more shelter from the blustery wind. We went ashore, looking for a bushwalk. Instead, we found a hut occupied by four big tough-looking blokes and a woman – deer stalkers who'd recently killed a deer. The blood-smeared deer was hanging upside down from a branch, its belly gaping vacantly. The hunters weren't happy to see us and dissuaded us from going bushwalking – probably for our own safety. We tried a couple of other tracks shown on our map, but they were too overgrown.

On Sunday, 28th February, we heard that there had been a very large earthquake (8.8 on the Richter scale) just off the coast of Valparaiso in Chile, killing 214 people. Tsunami warnings were issued for the entire Pacific. We heard that the water level rose four metres in Hiva Oa, an island in western French Polynesia. New Zealand took the warning seriously and moved people away from beaches and shore lines. Ian studied our position carefully. He reasoned that even if a three or four metre wave came in, we'd be okay. A rocky barrier lay to the east of us protecting us from the open sea. Ian let the anchor chain out a few more metres and took note of the time of high tide. When it arrived, the tide was about 150mm higher than normal. The anticipation had been more exciting than the reality.

We realised it was perfectly safe to sail the short distance down the coast to Port Adventure, as we'd planned. An easterly breeze took us south at a comfortable speed, but as the afternoon waned, so did the wind. We motored the last few miles, eyeing the various petrels and albatross sitting on the water as they warily eyed us. We tried photographing them to no avail. Just before six o'clock, we dropped anchor in Abrahams Bosom, where several power boats were already rafted up. *Osprey* would have shelter in the bay from yet another strong south-westerly which was forecast.

We received more rain, but no south-westerly. Instead, the next day the rain eased and, although the sky was still overcast, there was a light breeze from the north-east. As it wasn't raining and the Heron River entered at the head of the inlet, Ian decided he'd like to take the dinghy up the river. If we couldn't get into the bush on foot, we'd go an alternative way. It was 1st March, 2010.

After we had returned to Australia, I wrote the following for *The Mainsheet*, our cruising club magazine:

Up the Heron River

On Monday morning after breakfast, Ian decides to dinghy up the Heron River. We are at Stewart Island in Port Adventure, more than 47 degrees south. I reluctantly agree to go with him. I am in the middle of reading a good novel. In the end, I am pleased I go too.

Although the water is very clear down here, we find the light is such that we can't see much below the surface as we motor past the off-lying rocks, around to the mouth of the river. At one point, I see rock just under the surface as our bow passes. Fortunately, the propeller misses it. We stop once to clear kelp off the prop. The rain holds off, but the sky is very overcast.

The surface of the water is scattered with leaves, small sticks and kelp. As we move further up the river there is no more kelp, but bubbles of froth swirl with the currents. The water, thick with silt, is the brown of melted milk chocolate. The exposed banks are a dark chocolate under the bare, black, wet lower branches and trunks of the trees. Closer to the water, a quagmire of muddy sand supports a few birds, but nothing more. Some cormorants roost together in overhanging branches. We see a tern and several herons and can hear tuis.

The native bush overhead brings sensations reminiscent of my childhood – a sense of peace and tranquility that I found in the bush then, as well as the smell of wet dripping vegetation. The range of verdant greens seems more natural to me than the blue-greens of the Australian bush.

Pungas (tree ferns) push their fanned tops towards the light between other familiar trees I can no longer name. One tree gives the impression it might be a variety of pohutakawa with its small clusters of sodden scarlet blossoms. The colour of the dead fern-fronds stands out – a rusty red-brown. Another tree has reddish tips on its newest leaves.

Lichen hangs everywhere, like Christmas tinsel. Except for the palms, all the trees are festooned with this pale green lace. The rocks look as though they have a mustard green, short, loop-pile carpet draped over them. Wet-black supplejacks loop between the trees, reminding me of tinsel.

We reach the navigable end of the stream, where a tributary has brought rocks and boulders with it. The tide is fully out and we can take the dinghy no further. We turn off the motor and drift for a while, listening for birds. Soon, swirls of incoming tide start to drift us backwards. Ian starts the outboard again. We watch the currents rotate as the outflowing water meets the incoming tide.

As always, the return journey seems faster and, before long, the hunters' hut comes into view. We turn into the small bay and pull the dinghy up onto the beach. I step out and my insides curl as I look down at the pelvis and hind leg bones of a small deer. Elsewhere on the beach, I see a fish head and a fish carcass, partially chewed. I go up to the hut and read the instructions on the door. It appears that wild cats are also a problem pest here. There is a request for hunters to bait the traps for cats and rats before leaving the hut.

There is supposed to be a track from this beach north to Little Glory Bay. We find the beginning of the path and the signpost. Only a hundred metres into the bush, we lose the track already. I wait while Ian scouts further. He finds no more little red tags on the trees. We return to the beach.

There, we are puzzled to observe some rocks have emerged nearly a metre out of the water. Within a few minutes – perhaps fifteen – they are resubmerged to the level they were when we came in half an hour earlier. Tsunami variations in the water levels seem to be the obvious explanation. It is two days after the big earthquake in Chile. I wonder if there has been a further quake, and that is confirmed on the news the next morning.

To warm ourselves a little, we walk the length of the small beach and back to the dinghy. Despite being warmly dressed and in full waterproofs, we are cold from sitting so long in the dinghy and I am pleased to return to *Osprey* for a hot drink. Our expedition has taken two hours.

Jan outside the Parks & Wildlife hut

The cabin thermometer read nine degrees for a second morning when we climbed out of bed, and it was much colder outside. The temperature rose only a few degrees more during overcast days. We'd had enough of being cold. It was time to head north again. At first we thought we'd return to Oban, going into Half Moon Bay to refill our water tank and top up on diesel. When Ian listened to the forecast, he changed his mind. He decided to run for Port Chalmers (Dunedin) with the southerly wind that was coming that afternoon.

12 Returning North

'I've no regrets about turning back,' said Ian. 'Going further south to Pegasus Bay no longer seems attractive.'

'I agree. It's time to head back. Much as I've really enjoyed my visit to Stewart Island, it is just too cold and too far south to go on.'

The trip north from Paterson Inlet in Stewart Island was fast. We left in early March and it took us about twenty-eight hours to reach the Otago Yacht Club. For most of the way north we had one knot of current with us and we were scooting over the ground at between five and seven knots.

Ian misjudged our speed and we overshot Stirling Head, forcing us to motor-sail back hard into the wind and swell to make the Port Chalmers harbour entrance.

The tide was rising too, and we came up the channel into the harbour at six knots. The wind was gusting and when we were four hundred metres from the turn into the yacht club, the wind switched to a strong southerly. We had a frantic struggle to get the sails down and the engine on in time to turn into the yacht club. Fortunately, the yacht club supervisor and another man were on the pontoon to help us tie up, otherwise we might have overshot it with nowhere safe to go.

Otago Yacht Club

The supervisor complimented us at not shouting at each other.

'I was concentrating too hard to think of shouting at anyone,' Ian replied.

Early on in our cruising career, we'd been appalled at how many men shouted at their female partners when they were anchoring or coming in to a wharf. In Durban in the 1970s, we'd watched a much-practised couple who merely used hand signals to communicate from bow to cockpit. Thereafter, we'd vowed to follow their example.

We were given access to the yacht club facilities. That first hot shower is always bliss. The hot water cascaded through my hair and over my body – so good after days and days of shivering sponge-baths.

The club was about three kilometres from the city centre and not on a bus route, so we did a lot of walking to find supplies. Later we sought out the library and bookshops too. A ramble over the extensive botanical gardens was a joy. The plants were very different from those we were used to but the garden reminded me a little of the Botanical Gardens in Perth, where the hillside park also overlooks the city.

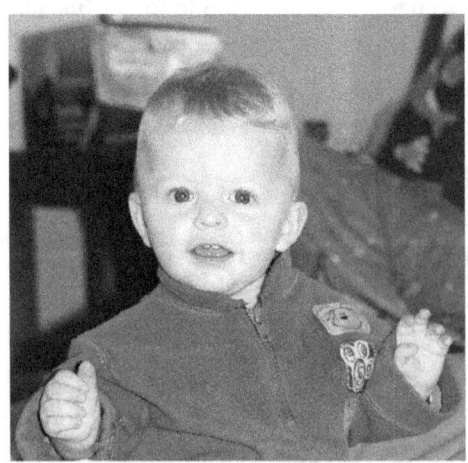

Our grandson, Tasman, at 16 months

I was keen to get back to Lake Macquarie for a visit to our grandson, Tasman. I knew the rest of my family and friends wouldn't forget me, but it was possible that a little boy of sixteen months might need a reminder about Nana.

Online, I found flights from Dunedin to Christchurch and then Christchurch to Sydney, returning a week later. A shuttle bus picked me up at the yacht club and I was away, leaving Ian on his own. I spent an uncomfortable night in the airport at Christchurch where I discovered I was already too old to sleep in airport seats so that I was ready for my early morning flight to Sydney. At Kingsford Smith, Heather and David picked me up from the train station at Turramurra. Tasman looked shy for a few moments.

'Hello Tasman. Do you remember me? Do you remember Nana?' I received a big grin. That was what I had come for.

⚓

'I'm pleased you're back. I was getting lonely,' Ian said on my return.

'I've missed you too,' I replied, hugging him tightly.

Ian had seen enough of Dunedin and he was ready to move on. I'd been there during my student days when I'd visited my best friend from school. It was March and still cold and I was pleased to be heading north again.

We picked up supplies the following day and left late on Wednesday afternoon for a two-nights-and-one-day hop up to Akaroa, sailing into the harbour early on Friday morning. From Taiaroa Head (Otago) to Akaroa Head, we experienced varied conditions, spending seventeen of thirty-four hours motoring and several hours sailing hard into the north-easterly sea breeze.

A couple of days later, it was autumn equinox - 22 March. There are often gales around the times of the equinox and the forecast predicted very strong winds from the north and north-west. We were anchored on a lee shore in French Bay, so we decided to head further up the harbour to find more shelter in Duvauchelle Bay. That night, we were awoken by the anchor alarm. Ian checked, cut the snubber he had put on the chain and let out more line. We stopped dragging. Thanks were owed to our new Rocna anchor; it had reset itself, the very quality for which Ian had bought it. In the morning, Ian estimated we had dragged about thirty metres before the anchor dug in again. We had made the right decision to change anchorage. In that bay, there was no other boat *Osprey* could have bumped into.

After we'd had a walk ashore, we listened to the afternoon forecast. Yet another strong southerly was on its way. Before the forecast had finished, the wind change arrived, gusting up the bay. Wind-waves of half a metre were soon pitching us violently. We needed to get to shelter.

In the rush to get the engine started and the anchor in, I didn't notice that Ian had already opened the engine cooling valves. He did this while I was busy turning on the switches for the anchor winch. Once again, I made the mistake of turning the valves off.

As we pushed against the wind and waves, smoke started to come from the engine compartment. We immediately realised what I'd done. Ian re-opened the valves and we motor-sailed, hoping the heavy plastic water box hadn't melted. The amount of smoke pouring out of the engine indicated we hadn't been so lucky. We anchored in Takamasa Bay, where, the next morning, Ian grumpily replaced the water box when the wind was calm. He explained it wasn't as simple as I'd thought to label those levers. Easier would have been for me to never touch the valves unless asked.

When we returned to French Bay we set about preparing to leave, doing domestic chores. As I was walking through the main street, a tourist climbed off a coach and sneezed in my direction. That had repercussions.

The weather pattern wasn't right to depart yet, so we hung about doing tasks on the boat for a couple more days and then I was glad we hadn't departed because I developed a bout of flu, with a sore throat, vomiting, fever and muscle weakness. After a day in bed, I consulted a doctor and was prescribed an antibiotic for the sinus infection.

It was no surprise that Ian also caught the flu too and was unwell for several days. Eventually, we were both well enough to leave, setting off on the 3rd April, Ian's sixty third birthday.

It was a very calm Saturday afternoon as we motored out of the harbour. There were lots of boats about – charter, fishing and private boats. As we left them behind, the wind came in and an hour later we were at sea, moving fast, seven and a half knots over the ground. That continued for some hours. The current was with us.

Although the wind dropped off early in the morning, it returned soon and, despite being changeable in force and direction, we had comfortable sailing all day Sunday and Monday. We motored most of Monday night, and early on Tuesday morning it was still calm when we rounded Cape Kidnappers into Hawkes Bay on the North Island.

Cape Kidnappers was named during a landfall there by HMS Endeavour in 1769. Local Maori tried to abduct a man from aboard the ship. Cook's journal states that Tiata, a Tahitian crew member, was in the water near *Endeavour* when a Maori fishing boat pulled alongside and dragged him aboard. Sailors on deck immediately opened fire on the fishing boat, killing two Maori and wounding a third. Tiata promptly jumped overboard and swam back to *Endeavour*, while the remaining Maori paddled their craft back to shore (Wikipedia).

We were heading for Napier and tied up mid-morning in the small marina at the Napier Yacht Club in the inner harbour. Rain settled in soon after our arrival and we cosied down for a quiet afternoon. Who wants anything else but to doze, read and eat after two nights of coastal sailing?

Napier had experienced a major earthquake in 1931. That and the subsequent fire killed 256 people and destroyed much of the city, making it New Zealand's worst disaster, except maybe for the much more recent Christchurch earthquake. The central city was rebuilt in art deco style, and the buildings still draw tourists today. It is a smallish, flat city of just over sixty thousand, making it possible to walk to most places in the business district.

While we were in Napier, I developed an infection which required more antibiotics. I was quite unwell for a couple of days. The sky stayed overcast, with occasional showers, and both of these factors put a dampener on our stay.

On Monday, we caught a bus across to Napier's twin city, Hastings, where we wandered about the shops for a couple of hours. By the time we returned to Napier, Ian had a sinus infection and was feeling poorly. He started antibiotics that evening and was looking worse the next morning before the medicine had time to take effect.

I was feeling positive about the *Sealegs* medication I'd bought in Akaroa to help with my seasickness, so I took it again. This time it made me quite dozy and I slept a lot, so that once again, Ian was single-handing. It was better than me heaving over a small wet bucket, though, when the smell of stale seawater alone is almost enough to make me sick.

Gisborne was our next port. We arrived there close to twenty-four hours after departing Napier. We tied up at a berth at the Sports Fishing Club, which provided showers and toilets for cruising yachties, and there was a public laundry only a short distance away.

I had never been to Gisborne before, so on Saturday we walked up to the visitor centre to get maps. I saw an advertisement for a local folk music club meeting that afternoon at the pub not far from the Sports Fishing Club, and we decided to go. They were a very friendly group singing in the bar. We joined in where we knew the words and it was a very enjoyable, relaxing afternoon.

The Titirangi Domain (Botanical Gardens) was quite close by, lying between the town and the beach and located on Kaiti Hill. We walked there on Sunday afternoon and found lovely gardens and walks. It was a beautiful sunny afternoon and we enjoyed ourselves, warm at last.

Later that day, we talked to an American couple on *Nine of Cups*. I started by asking them the meaning of their boat's name. They told us it is the name of the luckiest Tarot card. They had hired a car and offered to take us with them to the supermarket, which we gladly accepted.

Pip and Ros of *Isis* had invited us to visit them in Whakatane. We rang them to say we were in Gisborne and thinking of hiring a car to drive up. They told us we'd be welcome to come, and we booked a car for the following day. The route to Whakatane took us over a very scenic hill drive that winds down through a gorge and back up the other side. It was a lovely change to look at the countryside after spending so much time on the water.

There were roadworks around the gorge, which meant it was a slow trip, but who cared? We arrived in time for lunch. Pip had brought his mother down from Auckland to stay, which meant the guest room was occupied. Pip also admitted to us that he had serious cancer, which came as a shock to us. We ended up staying the night in the only motel room available – a posh room with its own spa bath. After booking in at the motel, we didn't get back there until bedtime, so the next morning we made sure we had a good soak in the spa. It was delightful to have some luxury after the fairly spartan way we live on *Osprey*.

After we had signed out of the motel, Pip took us for a tour of the local area, including to the river entrance to the yacht and fishing harbour. It was too narrow for our liking.

Isis was out of the water for some modifications because Pip was keen to do more sailing before his health failed too much, so we went to see what she looked like out of the water. Her hull was a very practical design for long distance cruising and her keel was long but not too deep.

The river port entrance at Whakatane

On Sunday afternoon, we set off back to Gisborne. If anything, the scenery on our return was more beautiful because the afternoon sun was shining through the autumn foliage – yellow, brown and red with some green remaining – along with the straw colour of the long roadside grass. The weather looked good to continue our run north the next day, so after returning the hire car, we packed up and motored out of our berth in the early afternoon. Unfortunately, the current was against us and we struggled to pass East Cape and then Cape Hicks and Cape Runaway, taking most of the night to make any headway. I managed to give Ian a four-hour break during the night, giving him a longer undisturbed sleep.

Eventually, the wind veered from a light north-easterly back into the west, which gave us a good run across the Bay of Plenty. I had spent two years as an English teacher in the Bay of Plenty before moving to Sydney in 1970. Although there is a good marina there at Mount Manganui, we didn't stop.

The current stepped in again as we started up the Coromandel Peninsula, so that once more we were moving over the ground at about walking pace. A small land bird flew into the cabin and then out again. It wasn't a sparrow, though about the same size. I wondered if perhaps it was some kind of swallow, yet I couldn't see any forked tail. It settled on the targa arch at the stern of the boat and stayed there for several hours, resting, but was gone when I looked next morning.

We rounded Cape Colville at the tip of the Coromandel Peninsula in the early morning and, with current against us, took until four o'clock to reach the anchorage at Waiheke Island, fifty nautical miles across the Hauraki Gulf. Knowing we would be in and around Auckland for the next week or so, we pulled the dinghy out of the forecabin onto the foredeck so that I could make up our double berth and return the dinette to its table function.

Arriving in the vicinity of Auckland seemed like a significant waypoint in our cruise and we marked it by having a 'rest' day. Waiheke is virtually a suburb of the city and many people commute by ferry to work in Auckland. The day was overcast, with occasional showers, so it was perfect to catch up on emails, bills and other necessities. Our internet connection was via a Vodafone modem to any local mobile phone tower.

Auckland Harbour from Waiheke Island

We heard from Jamie and his partner, Lisa, who were in the USA. They'd gone there to look for a cheap yacht after the Global Financial

Crisis and they had bought a run-down forty-five foot Morgan yacht. The yacht's engine was lying in a workshop in Jacksonville on the east coast of the Florida peninsula. After some months of repairs, they sailed *The Wild Goose* from the Gulf of Mexico to Jacksonville, where they repaired and installed the engine. They would soon set off on their voyage back to Australia via Puerto Rico, the San Blas Islands and Panama Canal.

We were enticed ashore the next morning at Waiheke Island by sunny weather. After locking up the dinghy we caught a succession of buses around the island, making several stops. First there was the ferry terminal and later the beach where we ate lunch.

We stepped off the bus again when we came to an 'anchorage' where the boats were sitting on the muddy sand, English-style. Never having been to Europe and having only seen such sights in photographs, we wanted to look more closely.

A shallow anchorage on Waiheke Island

Many of these boats seemed to be homes and I wondered what it would be like to have the tide recede, leaving a slightly muddy backyard behind, smelling of rotting fish and seaweed. How would you know when to take your dinghy ashore and when to walk?

We wandered around the main village on the island, which appeared to cater for tourists in the usual understated Kiwi way, and finally we made our way back to the dinghy.

Ian booked a berth for *Osprey* for several nights at the Westhaven Marina near central Auckland. We motored over there next morning, our noses already adjusting to the fumes of city traffic. Even though we knew this to be the biggest marina in the southern hemisphere, we were still shocked to see the size in reality. Each of the four sides is about a kilometre long. I had quite a long walk to find the office and amenities.

That evening, my niece, Karen, came for dinner. She is my oldest brother's elder daughter and our relationship has always been close. We chatted about family members and other matters until well into the evening.

Karen and her young daughter spent more time with us over the weekend. Eight year-old Morgan was keen to try sailing, but the sea breeze was very brisk and Karen and I were both mildly seasick. Morgan wasn't overly happy either, which was a pity for her.

We had intended to leave Auckland on Monday, but Ian was feeling unwell, and by Tuesday, moderate north-easterlies had set in, which would have made sailing conditions very uncomfortable. We chose to wait, and finally managed to leave the following Sunday.

The wind proved still to be too much 'on the nose' and when we couldn't clear Kawau Island, instead of tacking we chose to go in and anchor on the western shore. When we went ashore for a walk the next morning, we found there had once been a copper mine on the island. Finding the bay beside Mansion House Bay to be very sheltered, we moved *Osprey* there after breakfast.

The old Mansion House was a museum which had, so far, remained closed. When a ferry brought in a party of school children we realised the museum would be open for them, so we went ashore too. Inside the musty old house, there were many items of interest and we whiled away the morning.

Two couples from a local yacht were also at the museum and we began to chat. As yachties do, that afternoon we visited their boat and they visited ours. The following morning, we all went on a long walk over the island. Both couples belonged to a group of 'spirit-filled Christians'. They answered our questions and talked about their beliefs without becoming dogmatic and were also happy to drop the subject when we indicated we wanted to. This made for a much more pleasant morning than if they had preached hell-fire and damnation.

We'd spent nearly a week at Kawau and we really needed to move on. With the daytime forecast not favourable, we decided to motor overnight when the wind was calmer. We entered the channel leading into Whangarei Harbour before dawn and found a spot to anchor while we slept for a couple of hours before continuing the fifteen miles upriver to the yacht basin and a marina berth.

The passage upriver provided a different, and at moments, difficult experience. A large dredge that was motoring towards us was being passed by a large yacht on our side of the channel. We were pushed so close towards the shallow bank that the depth alarm was shrieking, though we didn't hit bottom. Then, as we were entering the marina, a fishing boat pulled out in front of us at the narrowest part of the channel and proceeded to make a three-point turn. In trying to avoid that vessel we came within one metre of a large catamaran in its berth. At times like these, my heart thumps, adrenalin pumps, and I become very anxious.

We found our berth in the marina and Ian steered in while I stood holding the shrouds and prepared to jump ashore with the bow line. Right where I was about to jump, there was a step screwed to the walkway and a bag sitting beside it. By the time I could get off, our bow was too close to the next boat, but someone appeared and took the line to stop *Osprey* in time. If there had been wind or current, we would have been in trouble. Sailing is usually a relaxing occupation and I dislike being faced with such unexpected stressors.

Being in another small city was a blessing. Everything was close by so that I could attend to domestic matters. That night, we went to bed soon after dinner and slept for eleven hours, undisturbed by the noise of traffic crossing the bridge just a couple of hundred metres upstream.

I went for a walk on my own after breakfast the next morning and lost track of the time. I wandered peacefully about the shops, selecting presents for family members and, at times, just browsing. By the time I returned, Ian had paid the marina fee, topped up the water tanks, brought in the electric cable and the washing line as well as washing up the dishes. With me back on board, he was ready to leave at the top of the tide. His plan was to stop overnight inside the channel again, so that we could leave early in the morning. However, the afternoon weather forecast brought a storm warning for the next day, indicating we'd be better off continuing on our way immediately.

We motored all the way to the Bay of Islands. The half-moon lit up the sky until 10pm, when it sank behind a bank of black clouds. We rounded the dark shape of Cape Brett just before midnight and crept across to Deep Cove, where we anchored before daylight. Ian found it unnerving to rely on electronic charts and GPS to find our way around the reefs and islets and I spent quite a lot of time on the bow squinting into the darkness ahead, trying to make out any dangers.

During the morning, the 'storm' arrived with heavy rain, and then the wind came. At first it was about thirty-five knots from the east, but it soon veered to the south-east and increased to fifty knots. Next, the wind swung south-west, with gusts up to 45 knots before easing. We might have had more shelter if we'd anchored in daylight and by the time we needed to move it was too late. We sat reading, with an ear out for any change indicating *Osprey* could be dragging her anchor. Ian usually wakes with an alteration of motion, so I slept soundly that night.

Before daylight, I awoke to hear Ian listening to the forecast. It was for more wind from the south-west. We were already on a lee shore and the tide was pushing the boat sideways. It was possible that the anchor would let go and she'd land on the beach.

'I don't like this situation, Ian.'

'Nor do I. I think we should move somewhere safer before the wind gets up again.'

'Let's do it, now.' I wanted to move before the wind rose any higher.

Ian brought the anchor in and I started the motor. Once again we were navigating electronically and in the dark. Ian had picked out a sheltered bay on the chart and to get there, we had to traverse a narrow pass. By the time we were halfway through, the sky had lightened enough for some vision of our whereabouts. A lot of our tension fell away. Finally we anchored, our stern pointing away from the beach where we were clear of any wind waves. *Osprey* was much safer there, and I wasn't going to be seasick.

After breakfast, we went ashore for a walk. The sky was still overcast and there were some spots of rain. We saw some scattered houses tucked away in the bush and the air had that fresh after-rain smell. It was cool and the wind gusts of thirty-five knots didn't touch us in that sheltered spot.

By Saturday morning, the sky had cleared. Our spirits always rise with a clear sky and we were happy to move again. This time we headed for the bay beside historic Russell, the original European settlement in New Zealand. We wandered through the old town that afternoon and ended up in one of the two museums. The museum included information about the treaty of Waitangi, its possibly deliberate mistranslation into Maori and how the wily old chief, Hone Heke, more than once cut down the flagpole erected by the Europeans he mistrusted. When I was a child, the day was peaceful, but in more recent times, since the realization that the Maori version was not an accurate translation, there has developed some discord between Maori and Pakeha (others; non Maori).

We moved yet again, closer to Opua this time. While we were at anchor Ian put on his wetsuit and scrubbed the hull in readiness for sailing home across the Tasman Sea.

Opua is the northernmost port for clearing Customs out of New Zealand. However, there is no town there – just the Customs Office, a general store, marina and the ferry terminal. For convenience, we decided to go into the marina before the next bout of heavy rain which was threatening. While Ian scrubbed out the dinghy and rinsed his wetsuit, both of which needed to be dry for stowing, I filled in the paperwork for our stay at the marina office and also checked out the Customs Office on the wharf. I found it was manned for only a couple of hours a day, but spare customs and immigration forms were available from a holder outside the office door.

The next few days in Opua were busy. I discovered that a small bus ran from the marina to the village of Paihea, eight kilometres distant. I made a visit there to top up our fresh provisions before the Tasman crossing.

A minor complication to our departure was that my good Kiwi mate, Annie, sent me some flowers for my birthday. The florist called and I arranged for them to be delivered to the marina office. There was a magnificent spray of white orchids and three exotic pink lilies with delicious yellow stamens. How she had spoilt me! We were booked with customs to leave the following day. I would have loved to take the flowers with me, but there was just nowhere for them to be safe on board. I enjoyed them for as long as possible and then left them at the marina office only minutes before our departure.

A couple of days later, on my birthday, I sang a snippet of the Beatles song and then asked Ian if he could still love me now that I was sixty-four. He was still sixty-three.

⚓

There was rain as we left the marina berth and half an hour later cloud descended over us, giving almost no visibility for a short time. It cleared as we proceeded towards the open sea and, by nightfall, the sky was almost cloudless with a full moon lighting our way.

We had plenty of wind to set sail, but the current was against us and two metres of swell was enough to hinder our progress and create a jerky motion. To me, this was as jarring as listening to jazz, which I dislike. We sailed all night to reach North Cape and, as we continued towards Cape Reinga, the current against us increased. The Three Kings Islands lie just north of Cape Reinga, the northern most point of mainland North Island. *Osprey* struggled to make headway and, after hours of sailing, we had to drop away and alter course to pass north of the islands. Ian called Radio Taupo to report our position. We were 1000nm from Newcastle.

When the wind dropped off we had to motor, otherwise the current took us backwards. Most of the time the easterly current was two or even two and a half knots, so that when we were sailing at six knots, we were making good over the ground at only three or four knots towards our destination.

Adverse current continued to slow us throughout our passage to Newcastle. It was also pushing us slightly northward. The last time we had travelled around the top of New Zealand had been in 1977 when we'd made course for Sydney in 25' *Caprice*. On that voyage, we'd made excellent speed, arriving home after ten and a half days, but in November, not June. I thought the current heading us was probably an arm of the East Australian Current, one of the huge eddies that sweeps east across the Tasman Sea. As the oceans warm, the EAC is extending its reach further southward. Maybe its eastern swirls and eddies are changing their extent and strength too?

On the afternoon of our fourth day out we were alarmed to hear a storm warning for sixty knots. The sky was ominously dark to the south-west and we took the precaution of reducing sail and lashing down potentially flying objects, but fortunately, the storm-strength winds didn't eventuate. Strong wind, rain and lightning did engulf us for an hour during which Ian managed to accidentally let go the furling line. The genoa whizzed out to its full extent, along with much crazy motion. Ian refused my offer to come out to help him refurl the sail. To do so on his own required him to head the boat into the wind, which resulted in the sail flogging so dramatically that the entire yacht was shaking and rattling. I feared the forestay would part and bring down the rig. Ian finally tamed the genoa, rolling most of it up. Fortunately, nothing had broken and I drifted back to sleep as we sailed on. *Osprey* is a tough little boat.

When we were about 600nm south of Norfolk Island two incidents suggested we were over the underwater Norfolk Ridge and that there were fish about. The first was when Ian spotted the lights of a vessel heading north-east. He tried calling them on the VHF radio. Eventually they answered but did not give their identity nor respond to Ian's question about whether they were towing. Ian gave them our position. We speculated that they might be illegally fishing or trawling. Secondly, that same afternoon, we were visited by a pod of large dolphins. They had white bellies and I thought they were common or bottlenose dolphins. The presence of dolphins also indicates that they are feeding on fish.

We both love it when dolphins come to visit. We rush up on deck and watch them. I even try to mimic their squeaks to get a response. While we had the motor running and *Osprey* was moving at over five knots, the mammals cavorted and raced each other in the bow wave. I tried to take some photos of them but, as usual, snapped nothing worthwhile.

After nearly a week of adverse current it eased up and for a couple of days we made excellent progress, covering 140nm on the fourth of June. This was possibly our best ever twenty-four hour run.

'We're out of the current at last. Now we can get a move on,' I said, ever hopeful. It wasn't to be. That current soon returned to dog us again. At sea, my mood rises and falls according to the weather and sailing conditions and I became despondent again.

By late afternoon, the wind and seas were rising. The barometer had peaked at 1018hPa, the wind was gusting to thirty-five knots and seas were breaking onto the boat. Ian took all sail off briefly while he reconfigured the sails, and we continued to move at five and half knots. Then we were running square before the wind at seven and a half knots, with a well-reefed genoa on the forestay out one side of the boat balanced on the other side by the storm jib hanked onto the inner forestay. The mainsail stayed furled up on the boom. The seas might have been rough, but we were moving along well.

The *Aries* had trouble controlling the tiller in such fierce conditions and during a wild swerve of our vessel, a wave slurped several litres of water inside, including over the HF radio. Neither of us had remembered to lower the clear plastic which usually protects the radios in bad weather. We mopped off as much water as we could from both the radio and the microphone. Later, Ian called Taupo Radio to report our position and alert them that the radio might cease communication because of the dousing. Next time we tried to give a position report, the radio failed to transmit although we could still receive and the VHF radio was still working. As we approached the coast its shorter range would still give us communication.

It occurred to Ian that the wind generator, which was whizzing furiously, could be interfering with the steering of the boat. He decided to muffle the generator and lo, the *Aries* began to steer without problems.

Later, Ian raised the trysail which further improved our course-keeping. We are always surprised at how much better that sail works than the triple-reefed mainsail.

The barometric pressure fell to 1014hPa by morning, and during the next day continued to plummet down to 996.5hPa – a massive drop in about thirty hours. The wind subsided, the current returned and we found – to our horror – we'd moved backwards over the ground by ten miles. We motored once again, lest we go further backwards.

As we proceeded, the swell increased to about seven metres, but they weren't breaking. Now we seemed to be going up and down more than forward, despite the gusty squalls which blew in to heel us when they caught the sails. We were probably over a sea mount which would interfere with the wave train and any current. I had taken *Sealegs* tablets nearly every day of the crossing and they had definitely moderated my tendency to vomit with every change of motion, but *mal de mer* ensnared me once again.

We unfurled a small amount of genoa on the forestay and used the storm jib on the inner stay and the trysail instead of the main. The downside of sailing *Osprey* as a cutter was the many lines crisscrossing the cockpit to access winches. There were two sets of sheet ropes from the two forward sails and also a pair of sheet ropes for the trysail, the foot of which was loose above the boom. We were also using running back stays, so the cockpit was turned into quite a spider's web with the lines for the *Aries* on the tiller too.

On two occasions, Ian recollects, he was perched on the upwind cockpit coaming when a wave broke on the stern. The motion threw him into the cockpit where the many lines created a sort of trampoline for him, preventing him from being injured.

Seasickness passes and on the night of our twelfth day at sea, the heavy swell disappeared too. I felt happy as we hurried along at seven knots for a few hours. Next morning, we were just under 200nm from Newcastle.

When the winds and sea moderated, I decided it was time to use a litre of fresh water to wash away the past couple of day's sweat. I had a hair wash in salt water, using conditioner to minimise the salt residue. How good it felt to be clean and in clean clothes. I also trimmed Ian's hair, to make him look tidier when we arrived at Newcastle.

It had been a difficult passage – sixteen days from Opua to Newcastle – with three gales of 30-35kts on the nose and more than normal contrary current. The cutter rig had enabled *Osprey* to sail to windward astonishingly well and Ian was very pleased he'd strengthened the deck to take an inner forestay. Despite the rough patches, the bouts of sea sickness and disappointment with the strong adverse current, I still enjoyed the voyage.

There is a sense of unity in the teamwork it takes to sail a yacht in adverse conditions, and I take pride in the fact that I can still help, even when seasick. Those rough times are countered by the sunrises and sunsets, the dolphins swimming in our bow wave, the beauty of uninhabited places, the joy of rapid friendship with fellow cruising yachties met en route, the intensity of the joint experience and satisfaction gained by sailing a small yacht across an ocean and bringing her and her crew safely back to land. Such experiences are part of the glue that binds Ian and me to our marriage and to our two sons who have both kept on sailing.

Before sunrise on our sixteenth day at sea, we crept past Nobbys Head and up the Hunter River into Port Newcastle. The Port Officer gave us a GPS position for a yellow mooring where we could tie up until we received further instructions from Customs.

In the low early morning light, we took a few minutes to locate the buoy, which was encrusted with barnacles and had no visible pick-up rope. Ian took one of our lines forward, along with some closed cell foam to lie on. With his belly over the bow and his feet tucked under the wire lifeline, he managed to loop the line through the ring on top of the buoy.

'Welcome to Australia,' he muttered sardonically when he returned to the cockpit.

At 9.00am, we received a call from Customs to proceed to a berth at the Newcastle Yacht Club marina. After motoring up and down looking for a 'Customs' sign, another yachtie directed us to an unmarked berth and helped us tie up.

Two quarantine officers and the marina manager arrived at the same time. Yes, we had to pay for the berth, even though we didn't intend to stay! The quarantine people had barely started when two burly customs officers came on board. Now six adults were squeezed into *Osprey's* tiny saloon. While one customs man brought out forms for Ian to fill in, the other started poking about in lockers. This was proving to be a far more intrusive procedure than we'd had in any port anywhere in thirty-six years of international cruising. The officer searched the bathroom, then pulled clean linen out of a bed-side locker where I'd also stowed the presents for family members that I'd wrapped for protection in pillow cases and tea towels. He pounced on these with glee, only to discover they weren't eligible for duty. I thought the officer seemed disappointed he hadn't caught us out. No wonder many yachties prefer to use Coffs Harbour as a customs port. Even Sydney Harbour Customs Officers had treated us more politely in the past.

We were greatly relieved when we were finally cleared into Australia. We intended to sail down to Lake Macquarie the same day, but having paid for a berth, we decided to make full use of the showers before leaving. We stayed under the hot water until our fingers were wrinkled.

Being mid-June, the days were short and it took all afternoon to motor-sail in light conditions to the Swansea channel. We passed through the Swansea Bridge just as the sun was setting, and then continued motoring through the rest of the channel and into Lake Macquarie.

Whilst the voyage along the east coast of New Zealand down to Stewart Island in the Southern Ocean was a wonderful experience, sailing home against so much current proved long and frustrating. Now we were keen to finish it – to put *Osprey* back on her mooring and be home. Sometimes I think it is as satisfying to arrive home from an adventure as it is to set out. But it doesn't take long to start wondering where our next adventure will be.

3 Lord Howe Island – again

My time for ocean cruising was coming to an end. After returning home from New Zealand I had a couple of falls. The psychologist in Sydney, who was treating me with passive neurofeedback for traumatic brain injury (TBI), advised me to stop sailing.

This saddened both Ian and me. The psychologist's testing showed that the concussion I had experienced soon after leaving Newcastle in December the previous year had created more damage. She spoke of possible permanent brain damage if I had any more serious knocks to my head. Ian talked of finding me a helmet to protect my head. For most of our marriage we had enjoyed the teamwork involved in sailing our various yachts to wonderfully isolated areas of the world and now that was going to cease.

Instead of concentrating on sailing, I focused on finishing writing Colin Kerby's biography, which I published in time for his 90th birthday on 11th November 2011. This was my first published book and it gave me both pride and confidence to continue with publishing our sailing stories. I continued being a regular contributor to *Cruising Helmsman* and also became more involved in my local writers' group.

Ian and I were excited about the prospect of a new grandchild. Lisa had become pregnant just before she and Jamie left the United States. They sailed *The Wild Goose* to Puerto Rico to visit a good friend, then to Cartagena in Columbia and on to the San Blas Islands on the Atlantic coast of Panama.

A local tribe there that had come to know Jamie well after several visits between 2006 and 2007, persuaded them to have a traditional San Blas wedding. The women dressed Lisa in their traditional wedding garb and the chief married them on 21st July, 2010. To this day they remain as committed to each other as they were that day more than ten years ago and have two beautiful children, Tane and Kira, who love living aboard *The Wild Goose* and, in 2019, are both learning to sail.

The Wild Goose came through the Panama Canal and started the long voyage across the Pacific. By the time they arrived in the Cook Islands, Lisa was eight months pregnant.

Jamie and Lisa are married, San Blas Islands

She flew from Rarotonga to Auckland and Jamie took on crew to help sail there. Our little grandson was born in Auckland late on 23rd December, 2010. In honour of his Kiwi birth, Jamie and Lisa named him Tane Mahuta, which means 'God of the Forest' in Maori.

When Tane was almost four months old, his parents crossed the Tasman Sea in *The Wild Goose*, clearing-in at Coffs Harbour in mid-April and then sailing down to Lake Macquarie. They stayed in the lake for the rest of the year. How wonderful it was to have family – especially grandchildren – close by.

Our younger son, David, Heather and their toddler, Tasman, were living in Newcastle and often visited, and I raised baby chickens for the delight of our tiny boys.

We had Tasman's third birthday in November, then Tane's first birthday two days before Christmas. We celebrated Christmas the same day, with a crowd at our place.

Ian and I wanted to have one last cruise together before I gave up ocean sailing altogether. The mild traumatic brain injuries I had sustained were interfering with my balance and making it more difficult to get over seasickness. The whole family decided to take a short cruise to Lord Howe Island.

On 27th December 2010, we set off from Swansea. Tasman stayed with his other grandmother, Jennifer, while on *The Wild Goose* David and

Heather sailed with Jamie, Lisa, baby Tane and two crew members. There was just Ian and me aided by Harry Aires on *Osprey*.

We had sailed to Lord Howe Island four times before. The first time, in 1973 in our Top Hat, *Caprice* (see *Two in a Top Hat*) when, due to atrocious weather, we were unable to land. Our voyages in *Realitas*, a Phantom 32, were in 1987, 1989 and 1992. Each of these times, we'd taken our boys to Lord Howe Island during the summer school holidays. (See *Crossings in Realitas*.) We all loved the island.

This was to be my final ocean voyage. I was sad, yet also relieved. Ian, in a wonderfully kind gesture, bought me a hockey helmet to protect my brain, but it made me feel top-heavy and altered my centre of gravity. I still bumped my head despite taking enormous care not to.

Ian and I departed Swansea within two hours after *The Wild Goose*. Jamie didn't have HF radio, and we lost contact on VHF that evening. The last we heard was that two of the crew members were fairly seasick. Apart from reminding Jamie to avoid the sea mounts, Ian didn't discuss our course with him at all. After all, he was now the more experienced sailor.

The wind stayed in the south-east. We sailed hard on the wind on the starboard tack for three and a half days. It was bumpy and uncomfortable and the humidity was a perspiring eighty percent. With spray flying all the time, we couldn't even open the forehatch. This wasn't the easy sail I'd anticipated, but I was going to enjoy Lord Howe Island. I always did.

Osprey required little attention for steering. Ian twitched the *Aries* lines occasionally, but mostly he kept the speed to a rate that the *Aries* could manage easily and which minimised discomfort for us.

We arrived off Erscotts Passage early on Friday morning, just before high tide. The lagoon is on the western side of the island and we hadn't been directed into Erscott's before. Clive Wilson, the Maritime Officer, sitting on the cliff top, used his radio to guide us in. When we arrived at our designated buoy, Ian was peeved that there was no mooring line attached. This was unexpected as we'd always found lines on the buoys on previous visits. Clive told us to attach a heavy nylon line. With the wind gusting up to thirty knots, it took Ian over an hour to get the mooring set up with a line from the buoy over the anchor roller and secured to the bow cleat. I struggled to hold the boat next to the mooring buoy while Ian worked on the bow.

We were puzzled and slightly worried that *The Wild Goose*, a forty-five footer, had not yet arrived. She had left just over an hour before us and, being a bigger vessel, should have arrived earlier. Lord Howe Radio had heard nothing from them. It niggled that both of our sons, their wives and one of our grandsons were on board. That was a lot of family members to lose if something happened to their boat. It was a relief when *The Wild Goose* turned up mid-afternoon.

The tide was too low for them to enter the lagoon then, so they were directed to anchor off the north of the island overnight. Their entry was scheduled for midday on New Year's Eve, when the tide would be high.

Jamie explained to us that they'd had trouble getting out of the Newcastle Bight. He hadn't been able to hold course as hard into the wind as *Osprey* could, and they'd had seas breaking into the cockpit. Ian thought maybe they'd had wind against current in the relatively shallow bight. Heather, with no ocean-going experience, had been particularly scared by the water in the cockpit and had already made up her mind that she would fly home.

The yachties already on the island had planned a party on shore that night and we were all keen to attend and eat a picnic dinner. It was a very pleasant evening, the temperature mild and the sky clear. As usual at yachties' gatherings, everyone was friendly. No one was loud, drunk or obnoxious. So sober were we all that we started back to our boats well before midnight, to prepare for bed. On the stroke of midnight, I added to the chorus of noise around the anchorage and the car horns on shore by blowing on our foghorn and Jamie's outline glowed red in the cockpit as he let off an old flare. Thus we said farewell to the old year and welcomed in 2012.

Our two crews decided to climb Mount Gower, but I declined and one of Jamie's crew stayed back too. I had preparations to make for New Year's dinner for all of us that evening. It was another noisy evening, drinking champagne toasts to Lord Howe Island and our being there. However, those who'd climbed Mount Gower were tired and headed off to *The Wild Goose* shortly after nine o'clock.

*The Mitchell family at the summit of Mt Gower:
Jamie, Tane, Lisa, Heather, Ian and David*

A couple of days later, I accompanied the crew of another visiting yacht, *Two Up*, on a walk up Mount Lidgbird. I had walked up the mountain on each of my visits to Lord Howe and once again enjoyed the native birds, vegetation and views from the Goat House Cave.

Before we had set off for Lord Howe, I had discussed with the editor of *Cruising Helmsman* the possibility of an article about the changes on the island since we'd last visited. She indicated that it was a good idea, but told me that she was leaving her job.

Apart from gathering material for the article, I had another mission – to photograph the remains of the RAAF Catalina which crashed on the island during World War II.

The home of the Catalina flying base during that time was at Rathmines where we now live. The historical society in our home area was preparing a book for publication about the role of the Catalina Flying Boats in the RAAF base here during and after World War II. I had promised them some photos of the crash site.

A Catalina crashed on this historic site in 1948

I submitted the following article to the new editor of *Cruising Helmsman* and it was published in Sep 2012.

Lord Howe – Jewel of the Tasman

Before dawn, not a light was to be seen apart from the occasional star, but with daylight, the peaks of Gower and Lidgbird were visible well over the horizon. The island was not so very far away, except that the current was taking us north, and we had been instructed to enter the lagoon through the South passage known locally as Erscotts Passage.

That current, which takes you away from the island as soon as you think you're near, hasn't changed, I thought to myself.

The tide was on the make and by the time we motor-sailed south-east against the current and wind, we were close to our desired location. On VHF channel 12, we conferred with Clive Wilson, once the Harbour Master and now the Maritime Officer. After a half-hour wait, Clive was in position on the cliffs to guide us into the lagoon.

Entry at Erscotts Passage

We found that the lagoon entry, via Erscotts Passage, was very complicated compared with our previous entries via the North Passage in *Realitas*, a Phantom 32 (See *Crossings in Realitas*). *Osprey* is a Joubert-designed Brolga 33, whose draft is approximately half a metre deeper.

We also had to cope with the wind blowing between twenty-five and thirty knots during this entry, with frequent gusts to thirty-five knots.

Still groggy from seasickness and dehydrated, I found manning the radio for instructions somewhat daunting. From the navigation station, I had to confirm via the radio that I had received the correct message and then relay each compass bearing to Ian in the cockpit. The instructions from Clive came every few seconds. We needed a portable VHF.

Previously, all moorings had their own ropes. The requirement to provide our own lines for attachment to the mooring buoy came as a surprise, although it may have been mentioned in the instructions that were sent to us when we booked our place in the lagoon. Finding strong enough lines and clearing the anchor off the bow roller before tying up took some time, but eventually we were secured. Clive was on standby all that time, ensuring the wind didn't blow us into danger.

Once moored, we could breathe more easily and take a look about. The lagoon was still crystal clear to its sandy bottom, the water that brilliant shade of aqua I remembered so well.

Our most recent visit previously had been in the summer of 1992-3, nineteen years earlier. I was eager to see what might have changed during those years and also hoped that the things I loved about the island had not altered.

The arrival of *The Wild Goose*

The Wild Goose, with the rest of our family and two crew members aboard, arrived a few hours after *Osprey*. Our son, Jamie, and his wife, Lisa, brought the 45' Morgan back from the USA in 2010-11. *The Wild Goose* was allocated the mooring next to ours. Like us, they had also had a fairly rough trip, and seasick crew do not steer as close a course as an *Aries* windvane. We would not make an ocean voyage without ours. Not only is it quiet and efficient, but it also doesn't argue with the skipper. We much prefer it to the autopilot, which is noisy and uses large amounts of electricity.

Island traffic

After catching up on sleep, we went ashore the next morning. The first thing we noticed was an increase in the number of cars and motor bikes.

Not only do the islanders have more cars for themselves, but they also hire cars and motor scooters to the tourists. The speed restriction is still 25kph. Once upon a time, tourists only had the choice of riding a bicycle or walking. Of course, many were still riding bicycles; but these were shinier and had gears. We also noticed that the roads were in better order, with fewer potholes.

The museum

We walked along Lagoon Road until we reached the Museum at the corner of Middle Beach Road. A beautiful new building had replaced the old one. It now includes a café, tourist information, two internet terminals, and a small lecture theatre. Books about the island and other items are for sale. The museum became our shore-side meeting place during our visit.

I was surprised to find flushing toilets for public use in the museum. All the new buildings seemed to have flushing toilets. I asked about composting toilets and was told that some had been introduced to guest houses, with mixed results. I also noted that there were many more water tanks evident and I didn't see one roof where rain water run-off was not directed into a tank. But collecting enough fresh water is only one side of the problem; contaminated ground water from leaking sullage is a downside and a continuing problem on Lord Howe Island, I was told.

Telecommunications

There are three public phone boxes on the island. One is solely for local calls; from the other two, we could make inexpensive calls to the mainland. Many of the locals have satellite phones and also use satellite to provide the internet for their homes, businesses and the guest houses. Otherwise, some islanders use the VHF radio to contact family members while out on business. We made use of the internet for email and checking weather forecasts.

Public facilities

Adjacent to the museum is the Aquatic Club building, a rather grand name for a shed where the islanders store their windsurfers and canoes. Next to the shed are two barbeques and picnic tables, as well as a garbage disposal unit – bins for waste, compost and recycling. The latter was another change. Previously, there was a tip at the southern end of the island, which smouldered and smoked. Now that area has become a waste sorting area.

On New Year's Eve, the visiting yachties took over the barbeque area for a gathering. It was a good way to meet the people whose boats were moored at the northern end of the lagoon. The islanders built their traditional huge bonfire beyond the wharf. The local teenage boys had a wonderful afternoon gathering flammable waste material and stacking it ready for the fire.

More coral

I love snorkelling. Three of us struggled into wetsuits and set off in the inflatable dinghy for Little Island where, in my opinion, exists the best coral snorkelling on the island. The thought that the beauty of this area might have been magnified in my mind in the intervening years concerned me. But no – if anything, it was more beautiful.

The warming of the oceans is bringing more coral to this spectacular island. Ian Hutton, the island's biologist, told us later that the coral in the lagoon has increased by 300%, much of it being brown coral, which attracts tropical fish, but is not yet as varied and colourful as the Little Island coral. At 22°C, the lagoon was definitely warmer than on previous visits, and only if I wanted to stay in the water for more than half an hour did I need a wetsuit.

Mountain Climbing

Our two crews hiked off to climb Mt Gower for the opening of 2012, with the exception of me,. They returned just before dark, sore, damp and tired, but exhilarated from their climb. Only one of them didn't make it to the top.

'There are more ropes installed now,' Ian told me. 'It's become an easier climb.'

I climbed to the goat house, as the cave halfway up the mountain is known, with some of the crew from *Two Up*. I wondered if there would be a tropicbird nesting there again. On the same nest was a red-tailed female, probably a descendant of the one I had photographed the previous time I had climbed up to the rocky overhang, which is as far as the public are permitted to climb. After that difficult climb, I was ready for the long hike to the northern end of the lagoon for a hot shower, to ease out my aching muscles.

Tropicbird on its nest, Mt Lidgbird

View of Lord Howe Lagoon

Amenities

The amenities block beside the wharf was erected about 1990. It had been renovated inside. The shower cubicles and toilets had been tiled and painted. The Lord Howe Radio Station, which used to occupy one room, seemed to have closed down. The washing machine had been moved into another room, with a large drying rack. The previous laundry room is now labelled 'Quarantine'. A key deposit of $20 gave us access to the showers and laundry. The toilets were unlocked.

Water is still available at a tap on the wharf and there is no physical restriction to how much you can use; just an admonition not to waste water. I think the wharf has been rebuilt, because it now has a dinghy landing area and I seem to remember climbing up a steel ladder last time.

Fresh produce

The *Island Trader* arrived with a cargo of stores, including fresh fruit and vegetables. It used to sail out of Yamba but now departs from Coffs Harbour. The tides were lower than normal at New Year and the vessel was unable to enter the lagoon for three days. Some fresh produce is flown in daily and, overall, freight costs to the island are very high. It cost me $6.00 to buy an iceberg lettuce and $2.50 for a small banana.

Beef cattle were once kept for slaughter, but Health and Safety have stretched out their long arm and stopped the local butchery. Likewise, local milk and eggs are no longer sold, but can be provided for family use. We saw several small herds of cattle, but no sheep. On the other hand, vegetable farming on a small scale was evident.

Hot water and electricity

As we walked about, we noticed many guest houses and homes had a solar hot water service, but we could see no solar panels for electricity. I enquired and found out that the island is part-way through a five-year plan to introduce sustainable electricity. The project began implementation in 2014. In the meantime, the diesel-powered generator opposite the community hall continued to run.

Seasonal changes

Walking along Lagoon Road, Ian and I expected to see terns nesting on the overhanging branches of Norfolk pines as on previous visits. Instead, to our surprise, fluffy grey tern chicks were sitting on some branches, but in many cases, the chicks had already fledged.

We questioned a local, who couldn't remember the season being later. Climate change was having a definite effect on the island. When we visited Middle Beach, hundreds of young mutton birds (fleshy-footed shearwaters) were practising flying from the cliff face.

'Did you notice, Ian,' I said, 'that when we walked through the palm forest, there were no squawks of young birds coming from the nests? Do you remember how noisy it used to be in the late afternoon as the chicks awaited their parents' return from their fishing?' Previously, nearly all the nests had contained chicks.

One outing advertised to visitors had been to walk across to the beach in the evening to eat a picnic meal, while watching the homecoming adult birds crash-land on the grass, then clumsily make their way on foot to their burrows under the trees to feed their raucous, hungry young.

Departure

We refilled our water tanks and, having washed the clothes and ourselves in fresh water, we were ready to depart. We had booked the mooring for seven nights, but eight had passed before tide and weather looked right for setting off. However, many of the islanders, including the Wilson family, are Seventh Day Adventists. Clive asked us to wait until Sunday morning to make our departure. He came out with his boat this time, instead of conning us from the cliff top. We ambled across the lagoon as Ian lashed the anchor back onto the bow roller and sealed off the naval pipe ready for sea.

I was sad to be leaving. I will probably never sail out to the island again and the alternative of flying and staying at a guest house is very expensive. Given that the island and the entire Lord Howe Ridge is a World Heritage site, the mooring fees of $28 per night for *Osprey A* (10m) and $37.40 for *The Wild Goose* (13.5m) are very reasonable. As well, we were charged an environmental levy of $36 per person and an administration fee per boat of $25. The $20 amenities key deposit was refunded on departure.

The island seemed just a little more manicured than it had been on our earlier visits, but the friendliness of the locals, the beauty of the beaches and lagoon, and the wonder of so much coral so close to the NSW coast still amazes me.

Paul, one of Jamie's crew, told me, 'I've fallen in love with Lord Howe Island. I definitely have to go back. I could live there for the rest of my life. I have travelled all over Australia, and seen nothing else like it.'

After only three days, David's wife, Heather, was keen to be home again. She was missing their toddler, Tasman and she flew out as soon as she could obtain a seat on the plane.

Jamie and Lisa stayed only five nights, but we had booked for seven, and ended up staying for ten nights in all. We snorkelled and relaxed. This was my last trip to Lord Howe and I wanted to make the most of it. We visited Little Island again, as well as the closer patches of coral.

Jan snorkelling (Photo by Ian Mitchell)

We had hoped to return home via Elizabeth and Middleton Reefs but by the time we left the weather was not suitable for going north and, instead, we turned for home.

At first, the sea was lumpy and I was glad I had taken *Sealegs*, although the drug made me drowsy. I was also putting the hockey helmet on my head every time I got up. I gained the impression that there was current about, a fairly usual situation close to the island.

As we moved away, the current eased, but it returned on Tuesday, this time with us, so that the sea calmed right down.

We were running before a south-easterly wind and the water temperature was 26.7°C – unbelievably warm. It made for a fabulous sailing day, but at this rate of warming, the world will reach two degrees above the year 2000's average temperature well before anyone is ready for it.

As we moved out of the current during the afternoon, the water temperature dropped to 25°C and our speed over the ground dropped off. The drop in the water temperature is a sure sign of a slowing current. On Wednesday, we found ourselves in the East Australian Current, taking us south-east. Again, the seas became rough and Ian was tacking the boat to try to make westing towards the coast. He was unable to lay a course for Swansea, so he decided to go into Port Stephens because a cold front was approaching.

It was quite late in the day when we finally arrived at Nelson Bay to find all the free moorings were occupied. We chose to motor around to Salamander Bay, where we had two attempts to anchor because the south-westerly arrived just then, gusting to thirty-five knots. The wind against tide created standing wind-waves of half a metre, but they soon settled down once the front had passed.

Ian encouraged me to go home by public transport from Salamander Bay, there being a bus stop not far from the anchorage. He was happy to sail back by himself, he said. So instead of sailing back into Swansea, my ocean-sailing life ended abruptly at Port Stephens.

To start with, I was relieved not to suffer from seasickness any more, but gradually I began to feel the loss of ocean-sailing in my life. This was contrasted with the knowledge that if I continued to get head injuries, I risked serious brain damage. I could also go blind from worsening macular degeneration if I subjected my eyes to more ocean glare.

My optometrist was convinced that long-term exposure to light reflection from water had created the damage. We learned that, despite wearing brimmed hats, light bounced back from the water and reflected off our spectacles. Ian has had to have multiple sun spots removed from his forehead and around his eyes, as have I, and we have each had one squamous cell carcinoma on our faces, requiring surgery.

Common sense told me it was time to stop. I would have liked to sell *Osprey* and travel to wild places by campervan or caravan instead, but Ian wasn't ready for that. He still isn't ready to stop listening to the allure of the ocean's song.

In 2018, I bought my campervan and decided I don't want Ian to sell *Osprey* if it makes him happy to keep her on her mooring near our house at Lake Macquarie. Ian enjoys going out to *Osprey* if the house gets him down, just to be on board on the water, doing routine maintenance. His psychological well-being depends on having a boat as an escape route from social obligations and to know he can take her out to sea if he wants. He often feels constrained by the house and agrees with Arthur Ransome who said, 'Houses are but badly built boats so firmly aground that you cannot think of moving them.'

By the time this book is published, Ian might have embarked on another voyage, hopefully shorter than his last on his own in 2013 to Tasmania for the birth of our granddaughter, Kira. He has already made one brief trip with me to the Gold Coast and a longer one to WA.

I still have itchy feet, and if I can't sail, then I want to travel Australia by campervan. Perhaps I still want more of that lovely companionship I have with Ian when we travel together to wild places. Ian still hankers to try solo-sailing and will try again in 2019. We didn't make it to the Kimberley this year, so maybe next winter or the one after. I still long to visit King Sound in the Kimberley.

If reading my memoirs of Ian's and my sailing adventures has increased your desire to cruise the oceans of our beautiful planet, you'd do best to follow the 1970s advice of the Pardeys: 'Go small. Go simple. Go now.' (*Cruising in Serrafin,* by Lin Pardey).

You don't know what the future holds. You don't know what health issues will arise for you and your family. Nor do you know the impact of climate change on our weather patterns. Climate change is already altering the intensity of storms and the warming of the oceans is changing currents as well as creating extremes of temperature and acidification of the waters, the latter affecting the growth of corals, crustaceans, shellfish, krill and, ultimately, the food chain.

In the five years since Hurricane Katrina in the USA, Australia has experienced Cyclones Larry, which flattened Innisfail in 2006 and Yasi, which landed just north of Hinchinbrook Island in 2011. The Philippines took a battering in 2016 and Texas in 2018. NSW is currently experiencing one of its most severe droughts since European settlement and an intense low pressure system has dumped nearly two metres of rainfall on Townsville and its hinterland. Fires in late winter and spring are no longer rare. Over sixty fires have raged in NSW this year (2019) and severe drought is widespread. This is just the beginning. Climate scientists predict an increase in severe weather events in the future.

My advice is: don't wait until you have finished with your career; don't wait until you have enough money to buy your dream yacht; go with what you can afford while you still can – go now. Let the ocean sing to *your* heart as it has to mine.

www.ingramcontent.com/pod-product-compliance
Lightning Source LLC
Chambersburg PA
CBHW071902290426
44110CB00013B/1246